Fractions

Power Math
Tutor Series

Author: Glenn Shepherd
Publication specialist
and illustrator: Kaytee Bell-Jarrett

F. E. Braswell Company, Inc.
P.O. Box 58734
Raleigh, North Carolina 27658
Phone (919) 878-8434
or (919) 878-9069
Fax (919) 878-9365

Published by F.E. Braswell
F.E. Braswell Co., Inc.
4910 Departure Drive
Raleigh, NC 27604

(919) 878-8434
(919) 878-9365 FAX

Ordering address:

F.E. Braswell Co., Inc.
PO Box 58734
Raleigh, NC 27658

Library of Congress Catalog Card Number 94-70737
International Standard Book Number (ISBN) 1-885120-02-8

Contents

This is a self-instructional manual that will help you learn on your own. This manual was designed with easy-to-follow directions and lots of help to maximize success. Here is what you do:

- **Read the text carefully.**
- **Answer the questions and solve the problems within the text.**
- **Check to make sure you can get the correct answers before going on.**
- **Some answers are on the next page. Answers to Practice Problems are at the end of the book.**
- **Go back and review a section if you have trouble with the problems.**
- **Work the additional Practice Problems at the end of the chapter. These problems cover the whole chapter. When you finish, check your answers with the ones at the end of the book.**
- **Do the word problems at the end of each chapter. Again, check your answers with the ones at the end of the book.**
- **Go through each chapter this way.**
- **When you complete the last chapter, do the Check Your Understanding problems. These problems cover the entire book.**
- **The answers to the Check Your Understanding are also at the end of the book.**

What are Fractions?

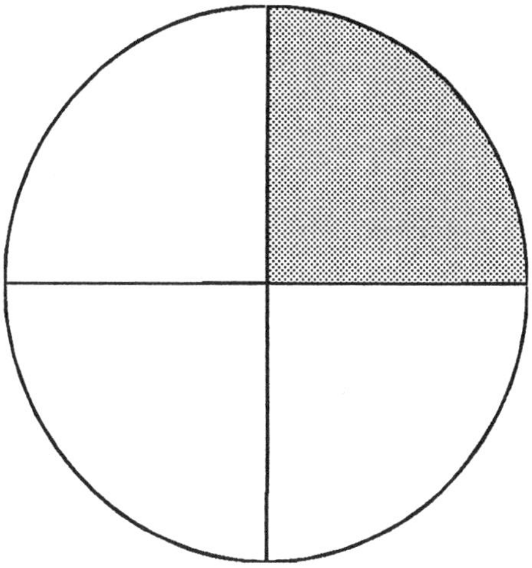

This is a whole pie. The shaded area is a part or **fraction** of the whole pie.

Following this lesson, you will be able to:

- Define fractions.
- Identify numerators and denominators.
- Identify proper (common) fractions, improper fractions, and mixed numbers.
- Change improper fractions to mixed numbers and mixed numbers to improper fractions.

Fractions are less than a whole. With numbers, 1 represents a whole. Fractions are less than 1.

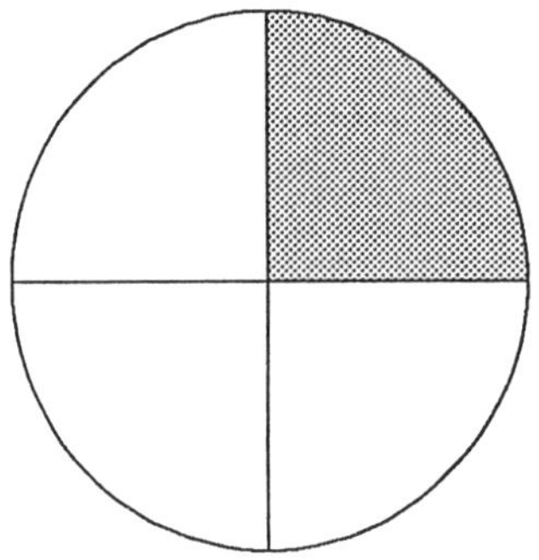

This pie is divided into 4 parts which make up the whole pie.

1 part out of **4** parts is shaded.

As a fraction, this shaded part can be written as:

$$\frac{1}{4}\ \begin{matrix}\text{numerator}\\ \text{denominator}\end{matrix}$$

The denominator is the number of total parts that the whole is divided into.

The numerator is the number of parts of the whole that is referred to.

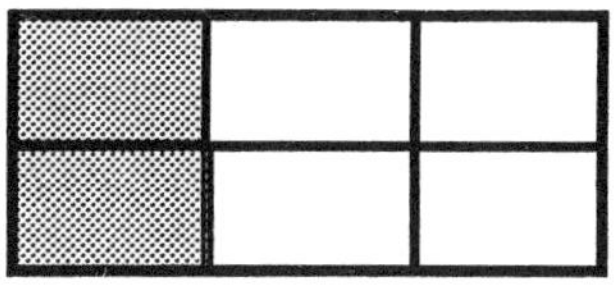 $= \frac{2}{6}$ This is a proper fraction.

When the numerator is less than the denominator, the fraction is called a proper fraction (sometimes called a common fraction).

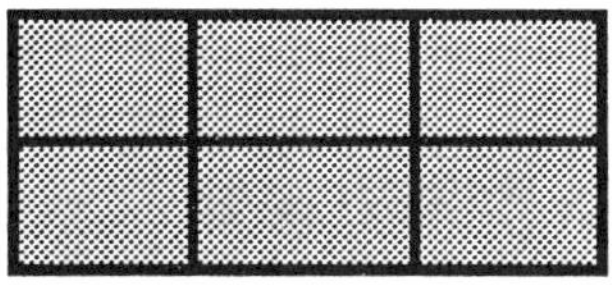 $= \frac{6}{6} = 1$

If the numerator is equal to the denominator, you have a whole or 1.

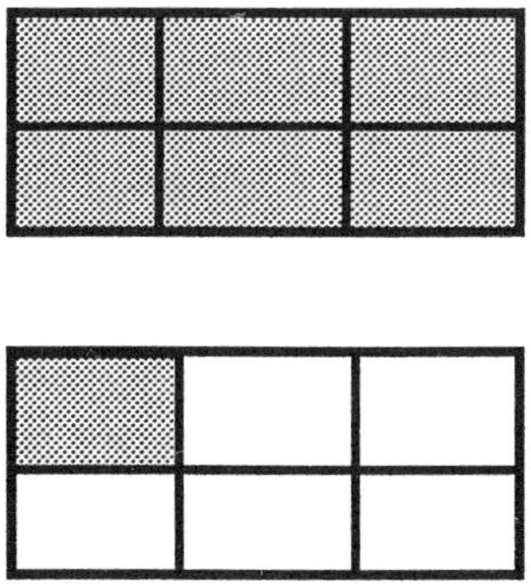 $= \frac{7}{6}$ This is an improper fraction.

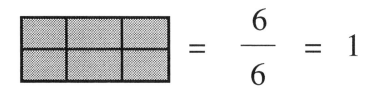

If the numerator is greater than the denominator, it is actually more than 1 and is not a proper fraction, so it is called an improper fraction.

$\frac{2}{6}$ This proper fraction can be reduced. This means there is a number that will divide into the numerator and denominator that will make each number less (or reduced).

Obviously 2 goes into both 2 and 6. So, $\frac{2}{6}$ reduces to $\frac{1}{3}$

Notice that $\frac{2}{6}$ is the same as $\frac{1}{3}$, so $\frac{2}{6} = \frac{1}{3}$

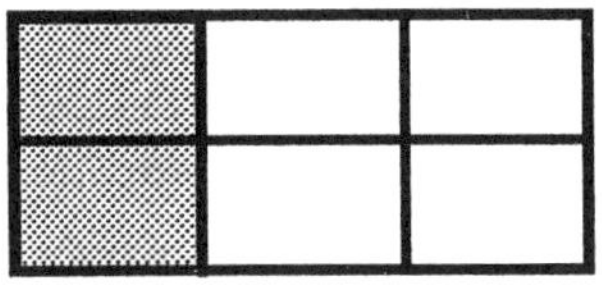

$\frac{1}{3}$

Sometimes, it is hard to find the number that will divide into both the numerator and denominator.

You try reducing these fractions:

1. $\frac{4}{28}$
2. $\frac{27}{51}$
3. $\frac{21}{105}$
4. $\frac{52}{117}$

Answers

1. $\frac{4}{28} = \frac{1}{7}$ (4 goes into both)

2. $\frac{27}{51} = \frac{9}{17}$ (3 goes into both)

3. $\frac{21}{105} = \frac{1}{5}$ (21 goes into both)

4. $\frac{52}{117} = \frac{4}{9}$ (13 goes into both)

In problem 3, you may have divided 3 or 7 into both the numerator and denominator. If so, you found out that the fraction had to be reduced again. It is alright to reduce two or three times, but if you can find the largest number that goes into both the numerator and denominator, then you won't have to reduce again.

Practice Problems: (Reduce)

1. $\frac{20}{45}$ 2. $\frac{6}{8}$ 3. $\frac{9}{27}$ 4. $\frac{30}{110}$ 5. $\frac{16}{32}$ 6. $\frac{24}{36}$

7. $\frac{50}{75}$ 8. $\frac{22}{55}$ 9. $\frac{70}{90}$ 10. $\frac{34}{56}$ 11. $\frac{35}{63}$ 12. $\frac{17}{136}$

13. $\frac{65}{117}$ 14. $\frac{21}{99}$ 15. $\frac{372}{492}$ 16. $\frac{57}{95}$ 17. $\frac{72}{168}$

18. $\frac{51}{119}$ 19. $\frac{18}{24}$ 20. $\frac{30}{65}$ 21. $\frac{88}{121}$ 22. $\frac{48}{96}$

23. $\frac{69}{115}$ 24. $\frac{64}{192}$ 25. $\frac{74}{112}$ 26. $\frac{15}{21}$ 27. $\frac{40}{60}$

28. $\frac{900}{1500}$ 29. $\frac{25}{75}$ 30. $\frac{27}{45}$ 31. $\frac{180}{245}$ 32. $\frac{63}{93}$

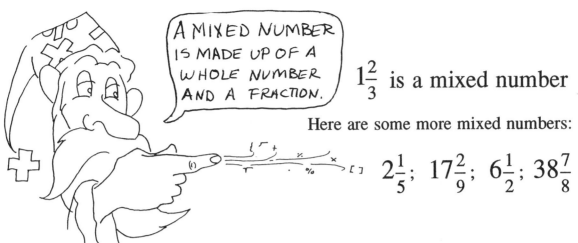

A MIXED NUMBER IS MADE UP OF A WHOLE NUMBER AND A FRACTION.

$1\frac{2}{3}$ is a mixed number

Here are some more mixed numbers:

$2\frac{1}{5}$; $17\frac{2}{9}$; $6\frac{1}{2}$; $38\frac{7}{8}$

Sometimes you need to change an improper fraction to a mixed number.

$$\frac{9}{5}$$

Just divide the denominator into the numerator.

Like this:

$$\begin{array}{r} 1 \\ 5\overline{)9} \\ -5 \\ \hline 4 \end{array}$$

1 → whole number

4 → left over (4 out of 5)

5 goes into 9 one time, so the whole number is 1.

Put the remainder (4) over the denominator (5).

$$\frac{9}{5} = 1\,\frac{4}{5}$$

Change the following improper fractions to mixed numbers:

1. $\frac{12}{11}$

2. $\frac{46}{7}$

3. $\frac{23}{5}$

4. $\frac{91}{13}$

Answers

1. $\frac{12}{11} = 1\frac{1}{11}$

$$\text{Look: } 11\overline{)12}\quad \begin{array}{r} 1 \\ 12 \\ -11 \\ \hline 1 \end{array}$$

2. $\frac{46}{7} = 6\frac{4}{7}$

$$\text{Look: } 7\overline{)46}\quad \begin{array}{r} 6 \\ 46 \\ -42 \\ \hline 4 \end{array}$$

3. $\frac{23}{5} = 4\frac{3}{5}$

$$\text{Look: } 5\overline{)23}\quad \begin{array}{r} 4 \\ 23 \\ -20 \\ \hline 3 \end{array}$$

4. $\frac{91}{13} = 7$

$$\text{Look: } 13\overline{)91}\quad \begin{array}{r} 7 \\ 91 \\ -91 \\ \hline 0 \end{array}$$

Notice that in problem 4 that there is no remainder, so you get the whole number 7.

If you had trouble with these, go back to page 6.

Practice Problems: (Change to mixed numbers and reduce)

1. $\frac{7}{6}$ 2. $\frac{11}{3}$ 3. $\frac{21}{5}$ 4. $\frac{14}{2}$ 5. $\frac{27}{7}$ 6. $\frac{40}{12}$ 7. $\frac{65}{18}$ 8. $\frac{31}{9}$

9. $\frac{52}{16}$ 10. $\frac{18}{5}$ 11. $\frac{24}{7}$ 12. $\frac{59}{13}$ 13. $\frac{512}{80}$ 14. $\frac{74}{18}$ 15. $\frac{19}{4}$

16. $\frac{37}{15}$ 17. $\frac{49}{20}$ 18. $\frac{13}{8}$ 19. $\frac{111}{40}$ 20. $\frac{66}{15}$ 21. $\frac{7}{2}$ 22. $\frac{8}{5}$

23. $\frac{34}{3}$ 24. $\frac{53}{13}$ 25. $\frac{215}{17}$ 26. $\frac{59}{30}$ 27. $\frac{85}{25}$ 28. $\frac{395}{82}$

29. $\frac{73}{9}$ 30. $\frac{46}{4}$ 31. $\frac{50}{30}$ 32. $\frac{100}{45}$ 33. $\frac{61}{9}$ 34. $\frac{88}{12}$

35. $\frac{69}{4}$ 36. $\frac{717}{150}$ 37. $\frac{283}{61}$ 38. $\frac{90}{15}$ 39. $\frac{73}{4}$ 40. $\frac{44}{7}$

41. $\frac{92}{6}$ 42. $\frac{26}{5}$ 43. $\frac{79}{18}$ 44. $\frac{61}{20}$ 45. $\frac{795}{115}$ 46. $\frac{94}{30}$

47. $\frac{125}{50}$ 48. $\frac{225}{100}$ 49. $\frac{996}{31}$ 50. $\frac{815}{49}$

Sometimes you may need to change a mixed number to an improper fraction.

$$3\frac{1}{4}$$

To change a mixed number to an improper fraction do this: Multiply the denominator times the whole number, then add the numerator to get the numerator of the improper fraction.

In the above example, you would multiple 4 x 3 (denominator times the whole number).

Then add this product (12) to the numerator (1).
12 + 1 = 13 The numerator of the improper fraction is 13 and the denominator stays the same (4).

$$3\frac{1}{4} = \frac{13}{4}$$

Change the following mixed numbers to improper fractions:

1. $1\frac{5}{6}$
2. $14\frac{3}{4}$
3. $8\frac{3}{8}$
4. 12

Answers

1. $1\frac{5}{6} = \frac{11}{6}$

6 x 1 = 6
6 + 5 = 11 (numerator of improper fraction)

2. $14\frac{3}{4} = \frac{59}{4}$

4 x 14 = 56
56 + 3 = 59 (numerator of improper fraction)

3. $8\frac{3}{8} = \frac{67}{8}$

8 x 8 = 64
64 + 3 = 67 (numerator of improper fraction)

4. $12 = \frac{12}{1}$

Any whole number can be written as an improper fraction by making the denominator to be 1.

Notice that in problem 4 you found that any whole number can be put in the numerator with a denominator of 1. Here are some more examples of this:

$$25 = \frac{25}{1}$$

$$136 = \frac{136}{1}$$

$$1 = \frac{1}{1}$$

If you had trouble, go back to page 9.

Practice Problems: (Change to improper fractions)

1. $1\frac{3}{8}$ 2. $4\frac{1}{5}$ 3. $5\frac{2}{3}$ 4. $7\frac{1}{4}$ 5. $3\frac{3}{7}$ 6. $8\frac{1}{2}$ 7. $5\frac{3}{4}$

8. $12\frac{1}{5}$ 9. $1\frac{9}{10}$ 10. $6\frac{3}{7}$ 11. $14\frac{1}{8}$ 12. $9\frac{3}{10}$ 13. $2\frac{11}{12}$

14. $30\frac{2}{3}$ 15. $8\frac{9}{13}$ 16. $4\frac{8}{9}$ 17. $5\frac{2}{5}$ 18. $42\frac{2}{5}$ 19. $7\frac{7}{20}$

20. $3\frac{4}{9}$ 21. $52\frac{1}{3}$ 22. $19\frac{3}{8}$ 23. $9\frac{1}{4}$ 24. $26\frac{11}{12}$

25. $10\frac{3}{8}$ 26. $4\frac{3}{4}$ 27. $13\frac{1}{7}$ 28. $9\frac{1}{2}$ 29. $14\frac{2}{3}$ 30. $6\frac{9}{14}$

31. $8\frac{7}{10}$ 32. $3\frac{2}{7}$ 33. $48\frac{3}{5}$ 34. $10\frac{3}{10}$ 35. $5\frac{21}{29}$ 36. $15\frac{1}{8}$

37. $1\frac{3}{11}$ 38. $2\frac{8}{17}$ 39. $8\frac{7}{8}$ 40. $4\frac{3}{5}$ 41. $9\frac{1}{16}$ 42. $51\frac{1}{9}$

43. $15\frac{2}{9}$ 44. $6\frac{3}{13}$ 45. $3\frac{9}{10}$ 46. $2\frac{1}{15}$ 47. $8\frac{1}{21}$ 48. $3\frac{7}{17}$

49. $23\frac{9}{22}$ 50. $72\frac{3}{8}$

Practice Problems

For the following problems, write down whether they are a proper fraction, improper fraction, or mixed number.

1. $\frac{2}{3}$ 2. $4\frac{1}{5}$ 3. $\frac{7}{8}$ 4. $\frac{17}{3}$ 5. $6\frac{4}{5}$

6. $\frac{9}{8}$ 7. $\frac{1}{12}$ 8. $\frac{25}{3}$ 9. $26\frac{1}{2}$ 10. $\frac{3}{7}$

Reduce the following:

11. $\frac{12}{15}$ 12. $\frac{6}{9}$ 13. $\frac{4}{18}$ 14. $\frac{35}{70}$ 15. $\frac{3}{45}$

16. $\frac{16}{24}$ 17. $\frac{48}{54}$ 18. $\frac{36}{51}$ 19. $\frac{9}{27}$ 20. $\frac{116}{120}$

Change the following to mixed numbers:

21. $\frac{13}{4}$ 22. $\frac{27}{5}$ 23. $\frac{41}{13}$ 24. $\frac{65}{9}$ 25. $\frac{19}{7}$

26. $\frac{49}{3}$ 27. $\frac{9}{7}$ 28. $\frac{21}{5}$ 29. $\frac{111}{20}$ 30. $\frac{7}{2}$

Change the following to improper fractions:

31. $5\frac{7}{8}$ 32. $3\frac{1}{2}$ 33. $9\frac{3}{5}$ 34. $3\frac{2}{3}$ 35. $16\frac{1}{6}$

36. $1\frac{6}{7}$ 37. $12\frac{5}{9}$ 38. $4\frac{3}{4}$ 39. $8\frac{1}{3}$ 40. $2\frac{5}{8}$

Practice Problems

For the following problems, write down whether they are a proper fraction, improper fraction, or mixed number.

1. $6\frac{4}{5}$ 2. $\frac{18}{7}$ 3. $\frac{27}{26}$ 4. $\frac{5}{9}$ 5. $\frac{3}{8}$

6. $\frac{16}{9}$ 7. $1\frac{1}{2}$ 8. $\frac{4}{7}$ 9. $12\frac{3}{4}$ 10. $\frac{29}{11}$

Reduce the following:

11. $\frac{18}{21}$ 12. $\frac{6}{9}$ 13. $\frac{25}{100}$ 14. $\frac{7}{21}$ 15. $\frac{16}{36}$

16. $\frac{120}{250}$ 17. $\frac{215}{320}$ 18. $\frac{17}{51}$ 19. $\frac{96}{104}$ 20. $\frac{146}{218}$

Change the following to mixed numbers:

21. $\frac{17}{9}$ 22. $\frac{61}{5}$ 23. $\frac{47}{6}$ 24. $\frac{14}{3}$ 25. $\frac{219}{6}$

26. $\frac{16}{7}$ 27. $\frac{37}{12}$ 28. $\frac{19}{2}$ 29. $\frac{53}{30}$ 30. $\frac{73}{20}$

Change the following to improper fractions:

31. $3\frac{5}{8}$ 32. $7\frac{1}{8}$ 33. $9\frac{1}{2}$ 34. $18\frac{3}{5}$ 35. $1\frac{7}{9}$

36. $45\frac{2}{3}$ 37. $13\frac{2}{5}$ 38. $4\frac{1}{10}$ 39. $2\frac{21}{100}$ 40. $5\frac{6}{35}$

[This page was intentionally left blank.]

Adding Fractions

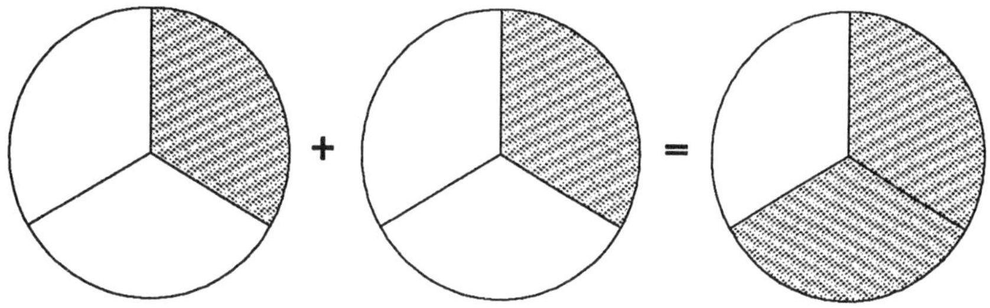

$$\frac{1}{3} + \frac{1}{3} = \frac{2}{3}$$

To add two fractions, the denominators must be alike (in common). Then you just add the numerators!

After this lesson, you should be able to:

- Add fractions with common denominators.
- Change fractions with different denominators to fractions with common denominators.
- Add fractions with uncommon denominators.
- Add mixed numbers.

It is easy to add fractions with common denominators. The trick is when the denominators are different.

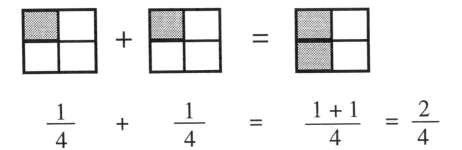

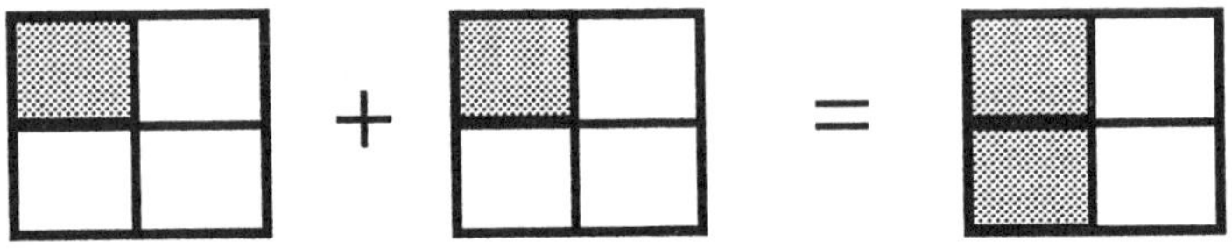

$$\frac{1}{4} + \frac{1}{4} = \frac{1+1}{4} = \frac{2}{4}$$

It is easy to see how you can add fractions when they have the same denominator. You are dividing the whole into the same number of total parts (the denominator).

So, all you have to do is add the parts (numerators) together!

Do these:

1. $\frac{3}{5} + \frac{1}{5}$

2. $\frac{5}{12} + \frac{1}{12}$

3. $\frac{2}{3}$

$+ \frac{2}{3}$

Answers

1. $\frac{3}{5} + \frac{1}{5} = \frac{3+1}{5} = \frac{4}{5}$

2. $\frac{5}{12} + \frac{1}{12} = \frac{5+1}{12} = \frac{6}{12} = \frac{1}{2}$

3. $\begin{array}{r} \frac{2}{3} \\ + \frac{2}{3} \\ \hline \end{array}$

$\frac{2+2}{3} = \frac{4}{3} = 1\frac{1}{3}$

In number 2, did you reduce $\frac{6}{12}$ to $\frac{1}{2}$?

In number 3, did you change $\frac{4}{3}$ to a mixed number?

You should reduce all of your answers and change improper fractions to mixed numbers.

Practice Problems

1. $\frac{3}{5} + \frac{1}{5}$

2. $\frac{4}{7} + \frac{2}{7}$

3. $\frac{8}{10} + \frac{7}{10}$

4. $\frac{3}{11} + \frac{5}{11}$

5. $\frac{12}{25} + \frac{11}{25}$

6. $\frac{7}{9} + \frac{4}{9}$

7. $\frac{2}{9} + \frac{5}{9}$

8. $\frac{3}{7} + \frac{6}{7}$

9. $\frac{4}{19} + \frac{8}{19}$

10. $\frac{17}{23} + \frac{5}{23}$

11. $\frac{14}{41} + \frac{23}{41}$

12. $\frac{5}{8} + \frac{3}{8}$

13. $\frac{1}{7} + \frac{5}{7}$

14. $\frac{17}{31} + \frac{6}{31}$

15. $\frac{19}{100} + \frac{29}{100}$

16. $\frac{21}{47} + \frac{19}{47}$

17. $\frac{3}{11} + \frac{2}{11}$

18. $\frac{2}{3} + \frac{2}{3}$

19. $\frac{15}{53} + \frac{31}{53}$

20. $\frac{7}{44} + \frac{29}{44}$

21. $\frac{91}{115} + \frac{23}{115}$

22. $\frac{21}{37} + \frac{18}{37}$

23. $\frac{17}{47} + \frac{8}{47}$

24. $\frac{9}{19} + \frac{7}{19}$

$$\frac{3}{4} \quad + \quad \frac{1}{8}$$

How could you add these fractions together? One block is divided into 4 parts and the other block is divided into 8 parts.

You can't add fourths to eighths unless you divide the whole into the same number of total parts (the denominators would then be alike).

To add fractions, the denominators must be the same.

To make the denominators the same, look at the numbers in the denominator.

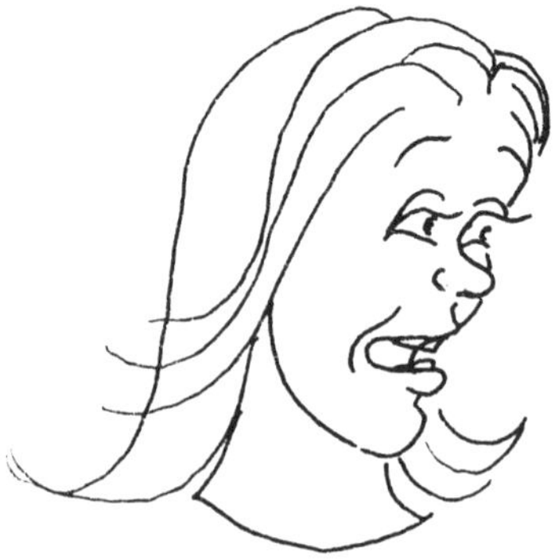

4 and 8

You have to change either one or both of these numbers, right?

Right. You must find a number that both 4 and 8 will divide into evenly.

There are several numbers you could have picked. The smallest number that both 4 and 8 will divide into is 8.

$$4\overline{)8}^{\,2} \qquad\qquad 8\overline{)8}^{\,1}$$

Even if you chose a larger number than 8, you could have solved the problem. You would just have to reduce the answer!

Now you can change both denominators to 8. One already has 8 as the denominator, so just change the 4 to 8. But how many eighths?

To change a fraction to a fraction with a different denominator you must find the new numerator.

$$\frac{3}{4} = \frac{?}{8}$$

How many 4's are in 8? $4\overline{)8}^{\,2}$

By dividing 4 into the new denominator you find out how many fourths are in one eighth (there are 2). So now you know:

$$\frac{1}{4} = \frac{2}{8}$$

But remember, you have 3 fourths. So how many eighths are in 3 fourths?

You would have 3 times as many eighths in 3 fourths as you have in one fourth.

$3 \times 2 = 6$ So, 6 is the new numerator!

When you have an addition problem with unlike denominators, write the new denominators and new numerators next to the old ones and then add. Like this:

$$\begin{array}{rcc} & \frac{3}{4} = & \frac{6}{8} \\ + & \frac{1}{8} = & \frac{1}{8} \\ \hline & & \frac{7}{8} \end{array}$$

Rule for finding the new numerator:

1. **Divide the smaller denominator into the larger denominator.**
2. **Multiply the answer by the numerator in the original fraction (this is the fraction with the smaller denominator).**
3. **The product is the new numerator!**

Find the new numerators of these fractions:

1. $\frac{2}{5} = \frac{?}{15}$

2. $\frac{4}{9} = \frac{?}{45}$

Answers

1. $\frac{2}{5} = \frac{2 \times 3}{15} = \frac{6}{15}$

2. $\frac{4}{9} = \frac{4 \times 5}{45} = \frac{20}{45}$

Now, you're ready to add fractions like these:

$$\begin{array}{rcl} \frac{4}{5} & = & \frac{}{35} \\ + \frac{2}{7} & = & \frac{}{35} \\ \hline \end{array}$$

First, find a common denominator and write it next to the original fraction. If you can't think of a common denominator, you can always get one by multiplying the two denominators together!

5 x 7 = 35

$$\begin{array}{rcl} \frac{4}{5} & = & \frac{28}{35} \\ + \frac{2}{7} & = & \frac{10}{35} \\ \hline \end{array}$$

Now, find the new numerator for each of the fractions.

$$\begin{array}{rcl} \frac{4}{5} & = & \frac{28}{35} \\ + \frac{2}{7} & = & \frac{10}{35} \\ \hline & & \frac{38}{35} = 1\frac{3}{35} \end{array}$$

Finally, just add numerators!

Remember to reduce your answer and change the improper fraction to a mixed number, if necessary.

Do these for practice:

1. $\dfrac{2}{9} + \dfrac{1}{4}$

2. $\dfrac{3}{7} + \dfrac{1}{3}$

3. $\dfrac{6}{11} + \dfrac{3}{8}$

4. $\dfrac{5}{13} + \dfrac{4}{7}$

Answers

1. $\frac{2}{9} = \frac{8}{36}$

$+ \frac{1}{4} = \frac{9}{36}$

$\frac{17}{36}$

2. $\frac{3}{7} = \frac{9}{21}$

$+ \frac{1}{3} = \frac{7}{21}$

$\frac{16}{21}$

3. $\frac{6}{11} = \frac{48}{88}$

$+ \frac{3}{8} = \frac{33}{88}$

$\frac{81}{88}$

4. $\frac{5}{13} = \frac{35}{91}$

$+ \frac{4}{7} = \frac{52}{91}$

$\frac{87}{91}$

Practice Problems

1. $\frac{1}{5} + \frac{2}{3}$
2. $\frac{2}{7} + \frac{1}{4}$
3. $\frac{7}{9} + \frac{1}{3}$
4. $\frac{3}{4} + \frac{3}{8}$
5. $\frac{6}{7} + \frac{1}{2}$
6. $\frac{1}{4} + \frac{1}{3}$
7. $\frac{2}{9} + \frac{1}{5}$
8. $\frac{7}{9} + \frac{1}{2}$
9. $\frac{5}{8} + \frac{2}{5}$
10. $\frac{1}{2} + \frac{3}{4}$
11. $\frac{7}{8} + \frac{1}{3}$
12. $\frac{7}{12} + \frac{3}{4}$
13. $\frac{4}{5} + \frac{2}{3}$
14. $\frac{1}{8} + \frac{1}{7}$
15. $\frac{2}{9} + \frac{1}{6}$
16. $\frac{2}{7} + \frac{1}{3}$
17. $\frac{4}{5} + \frac{1}{6}$
18. $\frac{3}{10} + \frac{2}{5}$
19. $\frac{3}{13} + \frac{2}{3}$
20. $\frac{1}{8} + \frac{2}{9}$
21. $\frac{7}{24} + \frac{3}{8}$
22. $\frac{9}{11} + \frac{1}{6}$
23. $\frac{2}{15} + \frac{3}{10}$
24. $\frac{7}{30} + \frac{3}{5}$
25. $\frac{13}{20} + \frac{1}{6}$
26. $\frac{4}{31} + \frac{2}{7}$
27. $\frac{14}{41} + \frac{1}{2}$
28. $\frac{7}{18} + \frac{1}{4}$
29. $\frac{6}{17} + \frac{7}{10}$
30. $\frac{8}{21} + \frac{1}{28}$

Practice Problems

1. $\frac{3}{7} + \frac{1}{5}$
2. $\frac{7}{12} + \frac{1}{8}$
3. $\frac{6}{19} + \frac{1}{2}$
4. $\frac{1}{4} + \frac{1}{2}$
5. $\frac{7}{8} + \frac{1}{6}$
6. $\frac{4}{13} + \frac{2}{5}$
7. $\frac{1}{9} + \frac{2}{3}$
8. $\frac{3}{4} + \frac{1}{3}$
9. $\frac{4}{21} + \frac{1}{7}$
10. $\frac{7}{20} + \frac{1}{4}$
11. $\frac{7}{20} + \frac{1}{4}$
12. $\frac{4}{9} + \frac{1}{2}$
13. $\frac{4}{5} + \frac{2}{3}$
14. $\frac{3}{8} + \frac{1}{6}$
15. $\frac{1}{11} + \frac{2}{9}$
16. $\frac{3}{10} + \frac{4}{5}$
17. $\frac{3}{7} + \frac{1}{8}$
18. $\frac{2}{15} + \frac{1}{3}$
19. $\frac{7}{13} + \frac{1}{4}$
20. $\frac{17}{30} + \frac{1}{5}$
21. $\frac{12}{43} + \frac{1}{3}$
22. $\frac{6}{7} + \frac{1}{9}$
23. $\frac{5}{8} + \frac{2}{3}$
24. $\frac{1}{16} + \frac{3}{8}$
25. $\frac{6}{17} + \frac{1}{4}$
26. $\frac{4}{9} + \frac{1}{5}$
27. $\frac{5}{24} + \frac{5}{18}$
28. $\frac{4}{15} + \frac{3}{10}$
29. $\frac{6}{7} + \frac{7}{11}$
30. $\frac{21}{50} + \frac{3}{20}$
31. $\frac{11}{12} + \frac{5}{8}$
32. $\frac{9}{15} + \frac{2}{3}$
33. $\frac{1}{2} + \frac{8}{9}$

Adding mixed numbers is not much different from adding proper fractions. You just add the fractions as you did before and then add the whole numbers separately.

$$\begin{array}{rl} 4\frac{2}{3} & = 4\frac{8}{12} \\ +\ 2\frac{3}{4} & = 2\frac{9}{12} \\ \hline & 6\frac{17}{12} = 7\frac{5}{12} \end{array}$$

Find the common denominator and move the whole numbers over.

Change the improper fraction to a mixed number and then add it to the whole number (6).

If you add a mixed number to a whole number, it is really easy! Look at these examples.

$$\begin{array}{r} 2\frac{1}{5} \\ +\ 3 \\ \hline 5\frac{1}{5} \end{array} \qquad \begin{array}{r} 6 \\ +\ \frac{2}{3} \\ \hline 6\frac{2}{3} \end{array} \qquad \begin{array}{r} 8 \\ +\ 2\frac{2}{7} \\ \hline 10\frac{2}{7} \end{array}$$

Now it's your turn:

1. $\begin{array}{r} 3\frac{1}{4} \\ +\ 2\frac{4}{5} \\ \hline \end{array}$

2. $\begin{array}{r} 9\frac{2}{3} \\ +\ 6 \\ \hline \end{array}$

3. $\begin{array}{r} 8\frac{1}{9} \\ +\ 3\frac{2}{5} \\ \hline \end{array}$

4. $\begin{array}{r} 7\frac{9}{10} \\ +\ 3\frac{2}{5} \\ \hline \end{array}$

Answers

1. $$\begin{array}{rcl} 3\frac{1}{4} & = & 3\frac{5}{20} \\ +2\frac{4}{5} & = & 2\frac{16}{20} \\ \hline & & 5\frac{21}{20} = 6\frac{1}{20} \end{array}$$

2. $$\begin{array}{r} 9\frac{2}{3} \\ +6 \\ \hline 15\frac{2}{3} \end{array}$$

3. $$\begin{array}{rcl} 8\frac{1}{9} & = & 8\frac{5}{45} \\ +3\frac{2}{5} & = & 3\frac{18}{45} \\ \hline & & 11\frac{23}{45} \end{array}$$

4. $$\begin{array}{rcl} 7\frac{9}{10} & = & 7\frac{9}{10} \\ +3\frac{2}{5} & = & 3\frac{4}{10} \\ \hline & & 10\frac{13}{10} = 11\frac{3}{10} \end{array}$$

Having any trouble? Remember:

1. Find a common denominator (you can always multiply the 2 denominators together and get a common denominator).
2. Move the whole number over.
3. Add numerators.
4. Reduce, if you can.
5. Change improper fractions to mixed numbers.

Practice Problems

1. $1\frac{1}{2} + 2\frac{1}{4}$

2. $6 + 2\frac{1}{3}$

3. $2\frac{1}{5} + 1\frac{1}{3}$

4. $3\frac{2}{9} + 1\frac{1}{6}$

5. $8\frac{2}{11} + 3\frac{3}{4}$

6. $2\frac{1}{3} + 3\frac{3}{4}$

7. $4\frac{1}{2} + 1\frac{1}{5}$

8. $6\frac{1}{9} + 9\frac{2}{5}$

9. $7\frac{1}{7} + 2\frac{1}{4}$

10. $3 + 8\frac{3}{16}$

11. $4\frac{2}{9} + 1\frac{7}{8}$

12. $10\frac{1}{3} + 4\frac{2}{5}$

Practice Problems

1. $4\frac{1}{5} + 1\frac{1}{7}$

2. $3\frac{1}{8} + 2\frac{2}{3}$

3. $4\frac{2}{7} + 2\frac{1}{9}$

4. $6\frac{1}{5} + 5\frac{2}{11}$

5. $4\frac{1}{15} + 2\frac{2}{5}$

6. $9\frac{2}{3} + 3\frac{1}{6}$

7. $5 + 6\frac{2}{5}$

8. $7\frac{1}{2} + 3\frac{3}{4}$

9. $2\frac{1}{9} + 6\frac{2}{3}$

10. $3\frac{3}{10} + 9$

11. $16\frac{1}{3} + 7\frac{2}{5}$

12. $4\frac{1}{19} + 2\frac{3}{4}$

13. $7\frac{1}{5} + 6\frac{3}{4}$

14. $6\frac{1}{2} + 9\frac{3}{5}$

15. $12\frac{1}{10} + 5\frac{2}{7}$

16. $42\frac{1}{3} + 17\frac{1}{4}$

17. $6\frac{3}{14} + 1\frac{1}{3}$

18. $9\frac{1}{12} + 4\frac{3}{4}$

19. $21\frac{1}{9} + 16\frac{2}{3}$

20. $3\frac{1}{15} + 6\frac{2}{3}$

21. $14\frac{1}{8} + 3\frac{3}{10}$

22. $6\frac{1}{5} + 3\frac{3}{8}$

23. $7 + 2\frac{1}{18}$

24. $5\frac{3}{11} + 2\frac{1}{4}$

25. $17\frac{1}{4} + 8\frac{1}{2}$

26. $35\frac{1}{6} + 11\frac{3}{4}$

27. $28 + 9\frac{2}{3}$

28. $14\frac{1}{3} + 6\frac{5}{6}$

29. $7\frac{9}{10} + 2\frac{4}{7}$

30. $5\frac{1}{17} + 3\frac{3}{10}$

31. $4\frac{2}{9} + 1\frac{7}{8}$

32. $10\frac{1}{3} + 4\frac{2}{5}$

33. $7\frac{9}{10} + 3\frac{1}{5}$

34. $6\frac{3}{7} + 4\frac{1}{4}$

35. $8\frac{1}{14} + 5\frac{1}{2}$

Practice Problems

Reduce answers:

1. $\frac{3}{8} + \frac{3}{8}$
2. $\frac{4}{5} + \frac{3}{5}$
3. $\frac{1}{4} + \frac{3}{4}$
4. $\frac{9}{16} + \frac{11}{16}$
5. $\frac{3}{11} + \frac{5}{11}$
6. $\frac{3}{5} + \frac{2}{3}$
7. $\frac{6}{7} + \frac{3}{4}$
8. $\frac{3}{8} + \frac{1}{4}$
9. $\frac{11}{12} + \frac{5}{8}$
10. $\frac{9}{13} + \frac{1}{2}$
11. $\frac{2}{15} + \frac{1}{4}$
12. $\frac{6}{11} + \frac{1}{3}$
13. $\frac{3}{16} + \frac{1}{32}$
14. $\frac{3}{20} + \frac{3}{10}$
15. $\frac{27}{100} + \frac{2}{5}$
16. $4 + 2\frac{1}{2}$
17. $9\frac{2}{3} + 5$
18. $6\frac{7}{8} + 7$
19. $14 + 6\frac{5}{8}$
20. $18\frac{1}{9} + 12$
21. $3\frac{1}{5} + 2\frac{1}{4}$
22. $5\frac{2}{3} + 1\frac{1}{2}$
23. $6\frac{1}{8} + 2\frac{3}{5}$
24. $9\frac{2}{9} + 4\frac{5}{6}$
25. $3\frac{1}{10} + 2\frac{2}{15}$
26. $14\frac{1}{4} + 10\frac{3}{8}$
27. $5\frac{1}{7} + 1\frac{1}{2}$
28. $6\frac{4}{9} + 2\frac{2}{3}$
29. $15\frac{1}{6} + 11\frac{1}{4}$
30. $21\frac{7}{8} + 9\frac{1}{6}$

Practice Problems

Reduce answers:

1. $\frac{1}{6} + \frac{5}{6}$

2. $\frac{11}{20} + \frac{3}{20}$

3. $\frac{8}{15} + \frac{4}{15}$

4. $\frac{6}{7} + \frac{3}{7}$

5. $\frac{9}{17} + \frac{3}{17}$

6. $\frac{2}{11} + \frac{3}{4}$

7. $\frac{1}{8} + \frac{4}{7}$

8. $\frac{12}{17} + \frac{4}{51}$

9. $\frac{8}{9} + \frac{1}{10}$

10. $\frac{18}{35} + \frac{3}{7}$

11. $\frac{3}{11} + \frac{4}{9}$

12. $\frac{9}{10} + \frac{1}{6}$

13. $\frac{19}{20} + \frac{2}{5}$

14. $\frac{16}{25} + \frac{1}{2}$

15. $\frac{21}{30} + \frac{2}{9}$

16. $8\frac{7}{9} + 9$

17. $14\frac{2}{3} + 7$

18. $10\frac{1}{2} + 7$

19. $56\frac{11}{12} + 19$

20. $41 + 19\frac{11}{25}$

21. $5\frac{3}{7} + 2\frac{1}{2}$

22. $6\frac{1}{5} + 2\frac{4}{9}$

23. $8\frac{1}{2} + 4\frac{5}{8}$

24. $26\frac{1}{3} + 12\frac{2}{9}$

25. $18\frac{1}{5} + 6\frac{1}{3}$

26. $12\frac{3}{10} + 4\frac{1}{4}$

27. $35\frac{2}{11} + 17\frac{1}{2}$

28. $16\frac{1}{2} + 13\frac{2}{5}$

29. $14\frac{6}{7} + 6\frac{1}{8}$

30. $56\frac{3}{35} + 19\frac{7}{10}$

Using Addition of Fractions

1. Judy walked $1\frac{2}{3}$ miles in the morning and walked $1\frac{1}{2}$ miles in the evening. How many miles did she walk that day?

2. A jeweler combined three diamonds to put into a ring. One diamond was $\frac{3}{4}$ carat, another was $\frac{3}{8}$ carat, and the last one was $1\frac{1}{2}$ carats. How many carats were in the ring?

3. The recipe calls for $1\frac{1}{3}$ cups of water and $\frac{1}{2}$ cup of milk mixed together. How many cups of the water and milk mixture would you use in the recipe?

4. The carpenter measured three pieces of molding that needed to be replaced. One piece measured $28\frac{3}{8}$ inches, and another measured $34\frac{1}{2}$ inches and the last one measured $44\frac{1}{4}$ inches. How many total inches of molding are there?

5. Don worked $9\frac{1}{2}$ hours on Saturday and $4\frac{3}{4}$ hours on Sunday. How many hours did he work over the weekend?

Using Addition of Fractions(cont.)

6. When Bobby was six years old, he was 45 $\frac{1}{4}$ inches tall. In the next two years he grew 3 $\frac{3}{8}$ inches. How tall was he at age eight?

7. Jane bought four packages of hamburger; one at 2 $\frac{1}{2}$ pounds, another at 3 $\frac{3}{4}$ pounds, another at 5 $\frac{1}{4}$ pounds, and 6 pounds. How many pounds of hamburger did she buy?

8. The weather station at the airport recorded 1 $\frac{1}{2}$ inches of rainfall in January, $\frac{1}{3}$ inch in February, and 1 $\frac{1}{4}$ inches in March. How much did it rain in the first three months of the year?

9. A woodworker bought a plank of cherry wood 10 $\frac{2}{3}$ feet long, a plank of oak 8 $\frac{3}{4}$ feet long, and a plank of mohogany 6 $\frac{1}{2}$ feet long. What is the total length of all three boards?

10. Tanya is writing a report. She wrote 10 $\frac{1}{2}$ pages one night, and 13 $\frac{2}{3}$ pages the next evening. How many pages is Tanya's report?

[This page was intentionally left blank.]

Subtracting Fractions

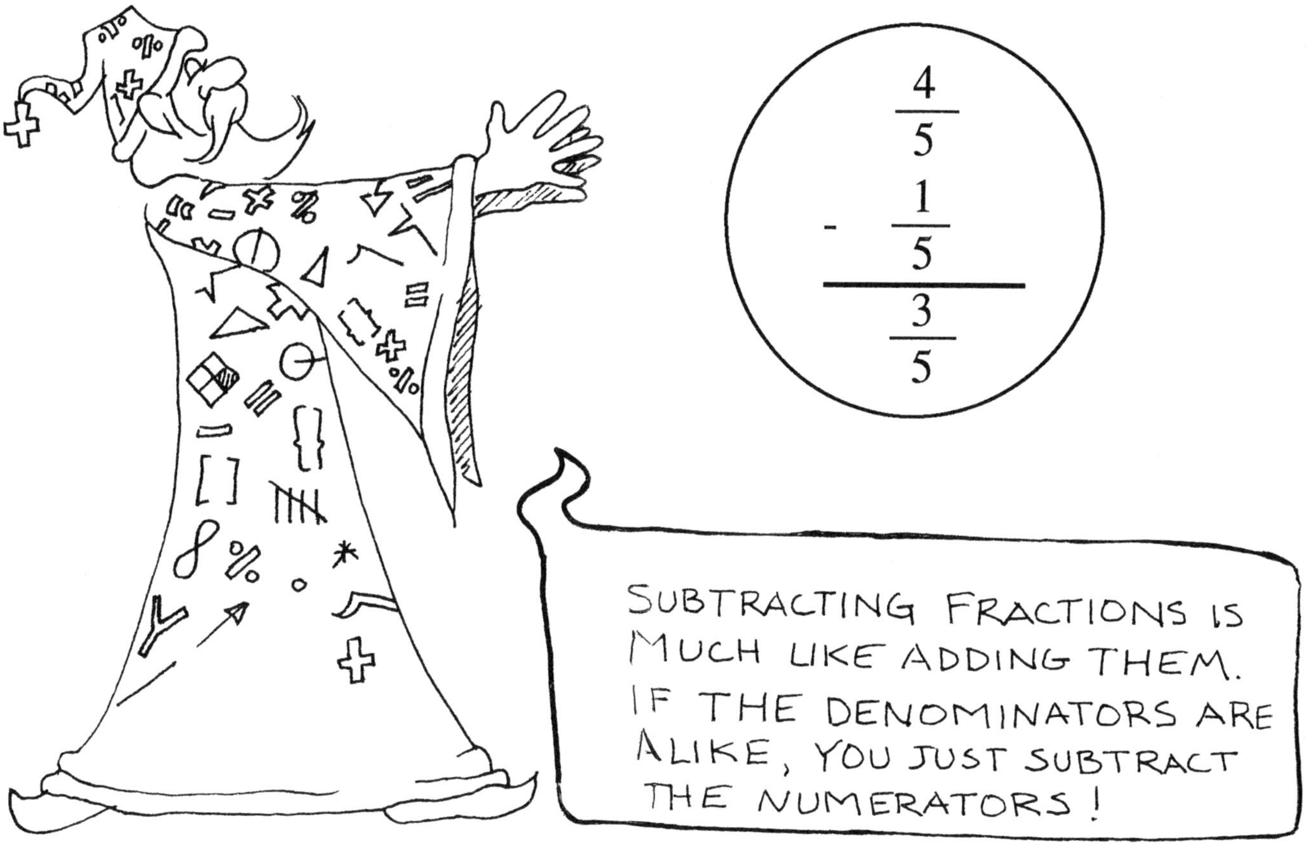

$$\frac{4}{5} - \frac{1}{5} = \frac{3}{5}$$

Following this lesson, you will be able to:

- Subtract fractions with common denominators.
- Subtract fractions with uncommon denominators.
- Subtract mixed numbers.

If you can add fractions, you'll be able to subtract fractions with no problem!

$$\begin{array}{r} \frac{9}{11} \\ -\ \frac{3}{11} \\ \hline \frac{6}{11} \end{array}$$

When denominators are the same, you subtract numerators and use the same denominator.

$$\begin{array}{rcl} \frac{3}{4} & = & \frac{9}{12} \\ -\ \frac{1}{3} & = & \frac{4}{12} \\ \hline & & \frac{5}{12} \end{array}$$

When denominators are different, change them to common denominators, then subtract numerators.

You try these:

1. $\begin{array}{r} \frac{7}{12} \\ -\ \frac{5}{12} \\ \hline \end{array}$

2. $\begin{array}{r} \frac{14}{19} \\ -\ \frac{6}{19} \\ \hline \end{array}$

3. $\begin{array}{r} \frac{7}{8} \\ -\ \frac{1}{4} \\ \hline \end{array}$

4. $\begin{array}{r} \frac{4}{5} \\ -\ \frac{1}{3} \\ \hline \end{array}$

Answers

1. $\dfrac{7}{12} - \dfrac{5}{12} = \dfrac{2}{12} = \dfrac{1}{6}$

2. $\dfrac{14}{19} - \dfrac{6}{19} = \dfrac{8}{19}$

Did you remember to reduce?

3. $\dfrac{7}{8} = \dfrac{7}{8}$, $-\dfrac{1}{4} = \dfrac{2}{8}$; $\dfrac{5}{8}$

4. $\dfrac{4}{5} = \dfrac{12}{15}$, $-\dfrac{1}{3} = \dfrac{5}{15}$; $\dfrac{7}{15}$

If you had trouble, review the lesson on addition of fractions.

Practice Problems

1. $\dfrac{3}{4} - \dfrac{1}{4}$
2. $\dfrac{5}{8} - \dfrac{3}{8}$
3. $\dfrac{8}{9} - \dfrac{2}{9}$
4. $\dfrac{8}{11} - \dfrac{5}{11}$
5. $\dfrac{6}{7} - \dfrac{2}{7}$
6. $\dfrac{5}{9} - \dfrac{1}{9}$
7. $\dfrac{11}{12} - \dfrac{7}{12}$
8. $\dfrac{12}{17} - \dfrac{4}{17}$
9. $\dfrac{5}{6} - \dfrac{1}{6}$
10. $\dfrac{3}{5} - \dfrac{2}{5}$
11. $\dfrac{21}{25} - \dfrac{11}{25}$
12. $\dfrac{7}{13} - \dfrac{3}{13}$
13. $\dfrac{14}{17} - \dfrac{6}{17}$
14. $\dfrac{9}{50} - \dfrac{3}{50}$
15. $\dfrac{21}{31} - \dfrac{13}{31}$
16. $\dfrac{7}{8} - \dfrac{5}{8}$
17. $\dfrac{8}{11} - \dfrac{7}{11}$
18. $\dfrac{14}{23} - \dfrac{8}{23}$
19. $\dfrac{6}{7} - \dfrac{1}{7}$
20. $\dfrac{33}{47} - \dfrac{11}{47}$

Practice Problems

1. $\frac{7}{8} - \frac{3}{4}$
2. $\frac{1}{4} - \frac{1}{7}$
3. $\frac{8}{9} - \frac{2}{3}$
4. $\frac{7}{11} - \frac{2}{5}$
5. $\frac{3}{4} - \frac{2}{9}$
6. $\frac{10}{11} - \frac{1}{3}$
7. $\frac{15}{16} - \frac{3}{8}$
8. $\frac{5}{6} - \frac{1}{4}$
9. $\frac{7}{10} - \frac{2}{5}$
10. $\frac{9}{13} - \frac{4}{9}$
11. $\frac{6}{7} - \frac{1}{3}$
12. $\frac{12}{25} - \frac{3}{10}$
13. $\frac{3}{7} - \frac{1}{5}$
14. $\frac{8}{13} - \frac{1}{2}$
15. $\frac{17}{18} - \frac{5}{9}$
16. $\frac{21}{28} - \frac{9}{14}$
17. $\frac{1}{2} - \frac{1}{3}$
18. $\frac{3}{4} - \frac{2}{5}$
19. $\frac{5}{8} - \frac{1}{4}$
20. $\frac{9}{16} - \frac{3}{8}$
21. $\frac{7}{8} - \frac{1}{2}$
22. $\frac{31}{40} - \frac{3}{10}$
23. $\frac{7}{9} - \frac{3}{5}$
24. $\frac{9}{10} - \frac{3}{4}$
25. $\frac{5}{7} - \frac{2}{9}$
26. $\frac{7}{12} - \frac{1}{4}$
27. $\frac{33}{61} - \frac{1}{2}$
28. $\frac{14}{17} - \frac{1}{5}$
29. $\frac{12}{23} - \frac{5}{46}$
30. $\frac{8}{9} - \frac{4}{7}$
31. $\frac{10}{12} - \frac{1}{3}$
32. $\frac{13}{24} - \frac{2}{6}$
33. $\frac{5}{8} - \frac{1}{2}$
34. $\frac{9}{32} - \frac{3}{16}$
35. $\frac{7}{8} - \frac{1}{4}$
36. $\frac{5}{10} - \frac{10}{50}$
37. $\frac{33}{43} - \frac{3}{13}$
38. $\frac{9}{10} - \frac{2}{9}$
39. $\frac{2}{7} - \frac{5}{21}$
40. $\frac{11}{22} - \frac{5}{12}$
41. $\frac{18}{37} - \frac{1}{3}$
42. $\frac{15}{23} - \frac{1}{4}$

Now, let's look at subtracting mixed numbers.

$$\begin{array}{rcl} 3\frac{5}{8} & = & 3\frac{5}{8} \\ -\ 1\frac{1}{4} & = & 1\frac{2}{8} \\ \hline & & 2\frac{3}{8} \end{array}$$

This problem is just like adding fractions, except that you subtract numerators.

There is one difference you need to know about. Look at this problem:

$$\begin{array}{rcl} 6\frac{1}{3} & = & 6\frac{7}{21} \\ -\ 3\frac{4}{7} & = & 3\frac{12}{21} \\ \hline \end{array}$$

In this problem, the top numerator (7) is smaller than the number to subtract (12). What do you do?

You have to borrow from the ones place, just like when you borrow in subtraction of whole numbers.

With fractions you always borrow 1. But remember, any number over itself (as a fraction) is equal to 1.

For instance,

$$\frac{21}{21} = 1$$

So when you borrow 1, change it to whatever denominator you are using in the subtraction problem.

In this problem, you only have $\frac{7}{21}$, but you need more.

So you borrow 1 or $\frac{21}{21}$.

$$\begin{array}{r} 6\frac{7}{21} \\ -\,3\frac{12}{21} \\ \hline \end{array} = \begin{array}{r} 5\,\frac{21+7}{21} \\ 3\,\frac{12}{21} \\ \hline \end{array} = \begin{array}{r} 5\,\frac{28}{21} \\ 3\,\frac{12}{21} \\ \hline 2\,\frac{16}{21} \end{array}$$

Understand? Remember, you borrow 1 from the 6 so you have to change the 6 to 5.

Look at one more:

$$\begin{array}{r} 12\frac{1}{4} \\ -\,7\frac{3}{5} \\ \hline \end{array} = \begin{array}{r} 12\,\frac{5}{20} \\ 7\,\frac{12}{20} \\ \hline \end{array} = \begin{array}{r} 11\,\frac{25}{20} \\ 7\,\frac{12}{20} \\ \hline 4\,\frac{13}{20} \end{array}$$

At first, you didn't have enough twentieths, so you borrowed 1 or $\frac{20}{20}$. Add the $\frac{20}{20}$ to $\frac{5}{20}$ and you have $\frac{25}{20}$, plenty enough to subtract.

You try these:

1. $$\begin{array}{r} 8\frac{1}{6} \\ -\,6\frac{4}{7} \\ \hline \end{array}$$

2. $$\begin{array}{r} 4\frac{2}{9} \\ -\,1\frac{5}{8} \\ \hline \end{array}$$

Answers

1.

$$\begin{array}{r} 8\frac{1}{6} \\ -\ 6\frac{4}{7} \\ \hline \end{array} = \begin{array}{r} 8\frac{7}{42} \\ 6\frac{24}{42} \\ \hline \end{array} = \begin{array}{r} 7\frac{49}{42} \\ 6\frac{24}{42} \\ \hline 1\frac{25}{42} \end{array}$$

2.

$$\begin{array}{r} 4\frac{2}{9} \\ -\ 1\frac{5}{8} \\ \hline \end{array} = \begin{array}{r} 4\frac{16}{72} \\ 1\frac{45}{72} \\ \hline \end{array} = \begin{array}{r} 3\frac{88}{72} \\ 1\frac{45}{72} \\ \hline 2\frac{43}{72} \end{array}$$

> **Did you:**
> 1. **Change to common denominators.**
> 2. **Borrow 1 from the whole number.**
> 3. **Subtract numerators.**

Easy right? Now, look at one more:

$$\begin{array}{r} 7 \\ -\ 3\frac{2}{5} \\ \hline \end{array} = \begin{array}{r} 6\frac{5}{5} \\ 3\frac{2}{5} \\ \hline 3\frac{3}{5} \end{array}$$

> **In this problem, the top number is a whole number. Just borrow 1, whatever the denomerator is on the bottom number.**

Practice Problems

1. $3\frac{1}{2} - 1\frac{1}{4}$

2. $7\frac{3}{4} - 4$

3. $5\frac{2}{3} - 2\frac{3}{8}$

4. $8\frac{1}{3} - 2\frac{1}{4}$

5. $4\frac{3}{8} - 1\frac{1}{9}$

6. $6\frac{1}{2} - 2\frac{3}{4}$

7. $2\frac{7}{9} - 1\frac{2}{5}$

8. $7\frac{7}{10} - 2\frac{3}{5}$

9. $6 - 2\frac{2}{3}$

10. $4\frac{1}{2} - 3\frac{5}{8}$

11. $6\frac{2}{3} - 5\frac{1}{2}$

12. $9\frac{3}{4} - 4\frac{1}{3}$

13. $12\frac{7}{8} - 7\frac{2}{5}$

14. $9\frac{1}{4} - 3\frac{5}{8}$

15. $5\frac{7}{9} - 1\frac{3}{5}$

16. $8 - 4\frac{1}{4}$

17. $14\frac{1}{8} - 7\frac{3}{4}$

18. $16\frac{1}{3} - 6\frac{5}{8}$

19. $8\frac{4}{5} - 3\frac{1}{2}$

20. $41\frac{1}{2} - 16\frac{5}{8}$

21. $15\frac{7}{9} - 11\frac{1}{3}$

22. $7\frac{1}{5} - 2\frac{5}{8}$

23. $4\frac{1}{9} - 1\frac{1}{3}$

24. $11\frac{2}{3} - 5\frac{1}{9}$

25. $20\frac{1}{6} - 8\frac{1}{3}$

26. $74\frac{1}{4} - 18\frac{4}{7}$

27. $15\frac{1}{8} - 7\frac{1}{12}$

28. $7\frac{3}{5} - 5\frac{2}{3}$

29. $41 - 17\frac{5}{8}$

30. $17\frac{2}{9} - 7\frac{4}{5}$

Practice Problems

1. $7\frac{1}{5} - 2\frac{1}{6}$
2. $5\frac{2}{5} - 1\frac{1}{4}$
3. $4\frac{3}{4} - 2\frac{1}{8}$
4. $6\frac{7}{8} - 3\frac{3}{5}$
5. $8\frac{7}{10} - 6\frac{1}{5}$
6. $4\frac{7}{9} - 1\frac{2}{3}$
7. $8 - 7\frac{1}{2}$
8. $6\frac{7}{8} - 3\frac{1}{3}$
9. $14\frac{1}{3} - 9\frac{3}{4}$
10. $21\frac{1}{8} - 6\frac{2}{5}$
11. $7\frac{6}{7} - 3\frac{2}{5}$
12. $5\frac{5}{9} - 3\frac{2}{7}$
13. $11\frac{2}{9} - 5\frac{1}{3}$
14. $6\frac{5}{8} - 3\frac{1}{4}$
15. $9\frac{5}{8} - 3\frac{3}{7}$
16. $17\frac{3}{5} - 12\frac{1}{3}$
17. $2 - 1\frac{2}{3}$
18. $7\frac{9}{10} - 4\frac{1}{6}$
19. $8\frac{3}{5} - 4\frac{5}{6}$
20. $16\frac{5}{11} - 12\frac{1}{4}$
21. $10\frac{1}{2} - 5\frac{3}{4}$
22. $5\frac{1}{3} - 2\frac{3}{8}$
23. $7\frac{1}{9} - 2\frac{1}{2}$
24. $18\frac{5}{6} - 6\frac{1}{4}$
25. $6\frac{4}{9} - 2\frac{2}{3}$
26. $8\frac{7}{8} - 3\frac{5}{6}$
27. $12\frac{7}{16} - 4\frac{1}{8}$
28. $19\frac{3}{4} - 6\frac{15}{16}$
29. $6\frac{1}{2} - 4\frac{7}{8}$
30. $35\frac{1}{6} - 19\frac{5}{8}$

Practice Problems

Reduce answers:

1. $\frac{9}{15} - \frac{2}{15}$

2. $\frac{8}{11} - \frac{3}{11}$

3. $\frac{12}{25} - \frac{7}{25}$

4. $\frac{12}{17} - \frac{11}{17}$

5. $\frac{18}{35} - \frac{9}{35}$

6. $\frac{5}{8} - \frac{1}{4}$

7. $\frac{3}{4} - \frac{1}{3}$

8. $\frac{7}{8} - \frac{2}{7}$

9. $\frac{6}{11} - \frac{1}{9}$

10. $\frac{15}{19} - \frac{3}{7}$

11. $\frac{7}{12} - \frac{1}{4}$

12. $\frac{7}{8} - \frac{2}{9}$

13. $\frac{6}{7} - \frac{1}{8}$

14. $\frac{11}{18} - \frac{5}{12}$

15. $\frac{8}{9} - \frac{11}{15}$

16. $6\frac{2}{3} - 2\frac{1}{12}$

17. $9\frac{9}{10} - 3\frac{3}{5}$

18. $5\frac{1}{2} - 2\frac{4}{13}$

19. $12\frac{7}{8} - 9\frac{1}{6}$

20. $31\frac{5}{9} - 14\frac{2}{5}$

21. $65\frac{12}{15} - 17\frac{7}{30}$

22. $8\frac{6}{11} - 3\frac{1}{4}$

23. $10\frac{19}{20} - 6\frac{3}{10}$

24. $5\frac{11}{12} - 1\frac{5}{8}$

25. $44\frac{3}{4} - 18\frac{7}{12}$

26. $2\frac{1}{4} - 1\frac{3}{8}$

27. $8\frac{3}{5} - 6\frac{7}{8}$

28. $24\frac{1}{5} - 16\frac{3}{7}$

29. $4 - 1\frac{3}{8}$

30. $18 - 7\frac{4}{10}$

Practice Problems

Reduce answers:

1. $\frac{6}{13} - \frac{4}{13}$
2. $\frac{18}{29} - \frac{7}{29}$
3. $\frac{14}{35} - \frac{8}{35}$
4. $\frac{28}{33} - \frac{10}{33}$
5. $\frac{16}{21} - \frac{11}{21}$
6. $\frac{6}{11} - \frac{1}{2}$
7. $\frac{12}{13} - \frac{2}{5}$
8. $\frac{10}{21} - \frac{3}{7}$
9. $\frac{5}{6} - \frac{1}{9}$
10. $\frac{6}{15} - \frac{1}{3}$
11. $\frac{15}{34} - \frac{5}{17}$
12. $\frac{17}{20} - \frac{4}{7}$
13. $\frac{9}{10} - \frac{5}{8}$
14. $\frac{6}{7} - \frac{1}{12}$
15. $\frac{15}{16} - \frac{11}{64}$
16. $9\frac{1}{3} - 3\frac{1}{4}$
17. $4\frac{7}{8} - 2\frac{3}{5}$
18. $5\frac{4}{9} - 4\frac{1}{7}$
19. $8\frac{6}{7} - 1\frac{1}{3}$
20. $28\frac{9}{17} - 17\frac{7}{34}$
21. $7\frac{1}{4} - 2\frac{1}{3}$
22. $9\frac{1}{2} - 7\frac{7}{8}$
23. $14\frac{1}{5} - 8\frac{3}{5}$
24. $16\frac{2}{3} - 8\frac{7}{8}$
25. $5\frac{1}{7} - 1\frac{5}{9}$
26. $16\frac{1}{9} - 15\frac{7}{12}$
27. $47\frac{3}{5} - 28\frac{7}{11}$
28. $18 - 4\frac{2}{3}$
29. $66 - 19\frac{3}{7}$
30. $75 - 48\frac{5}{12}$

Using Subtraction of Fractions

1. Before beginning her diet, Sally weighed 142 pounds. After one year of dieting, she lost $18\frac{1}{2}$ pounds. How much weight did she weigh after one year of dieting?

2. John filled the $15\frac{1}{2}$ gallon gas tank in his car, before leaving on a trip. When he returned from the trip, it took $12\frac{3}{4}$ gallons to fill up his tank. How many gallons of gas did John use?

3. The carpenter is building a deck and needs a $46\frac{3}{8}$ inch board. He cuts this piece from an 8 foot board. How many inches of board are left over? (Remember, there are 12 inches in a foot.)

4. A 5 gallon water tank has $4\frac{1}{2}$ gallons of water in it. Paul takes $1\frac{3}{4}$ gallons of water from the tank. How much water is left?

5. Diane bought $5\frac{1}{4}$ yards of material and used $3\frac{1}{2}$ yards to make a dress. How much material did she have left?

6. LaToya and her friends baked 4 dozen chocolate chip cookies and ate $1\frac{3}{4}$ dozen that same day. How many dozen cookies were left?

7. Bob drove to the bank and measured the distance at $2\frac{1}{2}$ miles. He then drove to work and figured his total mileage at 8 miles. How many miles is it from the bank to Bob's work?

8. A 4 inch strip of ribbon is cut from a $92\frac{1}{2}$ inch strip of ribbon. How much ribbon is left?

9. Darryl had a full gallon of paint and spilled $\frac{3}{8}$ gallon of paint before he got started. How much paint did he have left?

10. Carl bought 5 pounds of shrimp. He and his family ate $3\frac{1}{4}$ pounds of shrimp during dinner. How much shrimp was left?

Multiplying Fractions

$$\frac{1}{5} \times \frac{2}{3}$$

Multiplying fractions is actually simpler than adding or subtracting fractions. Why? Because you don't need common denominators?

After this lesson, you will be able to:

- Multiply two proper fractions.
- Multiply a whole number and a proper fraction.
- Multiply a mixed number and a proper fraction.
- Multiply two mixed numbers.

In order to multiply fractions, you need to know how to multiply whole numbers and how to change a mixed number to an improper fraction.

To multiply two fractions, just multiply the numerator times the other numerator to get the new numerator. Then multiply the denominator times the other denominator to get the new denominator.

$$\frac{1}{5} \times \frac{2}{3} = \frac{1 \times 2}{5 \times 3} = \frac{2}{15}$$

$$\frac{3}{8} \times \frac{2}{3} = \frac{3 \times 2}{8 \times 3} = \frac{6}{24} = \frac{1}{4}$$

If the number is a whole number, remember that the denominator of a whole number is 1.

$$6 \times \frac{2}{9} = \frac{6}{1} \times \frac{2}{9} = \frac{6 \times 2}{1 \times 9} = \frac{12}{9} = 1\frac{3}{9} = 1\frac{1}{3}$$

REMEMBER TO CHANGE THE IMPROPER FRACTION TO A MIXED NUMBER AND REDUCE YOUR ANSWER!

You do these:

1. $\frac{3}{7} \times \frac{2}{3}$ 2. $\frac{7}{11} \times \frac{5}{9}$

3. $4 \times \frac{1}{3}$ 4. $\frac{7}{8} \times 6$

Answers

1. $\frac{3}{7} \times \frac{2}{3} = \frac{3 \times 2}{7 \times 3} = \frac{6}{21} = \frac{2}{7}$

2. $\frac{7}{11} \times \frac{5}{9} = \frac{7 \times 5}{11 \times 9} = \frac{35}{99}$

3. $4 \times \frac{1}{3} = \frac{4}{1} \times \frac{1}{3} = \frac{4 \times 1}{1 \times 3} = \frac{4}{3} = 1\frac{1}{3}$

4. $\frac{7}{8} \times 6 = \frac{7}{8} \times \frac{6}{1} = \frac{7 \times 6}{8 \times 1} = \frac{42}{8} = 5\frac{2}{8} = 5\frac{1}{4}$

If you had trouble, go back to page 46.
If you forgot how to reduce, see page 4.
If you forgot how to change an improper fraction to a mixed number, see page 6.

Practice Problems

1. $\frac{2}{3} \times \frac{1}{5}$
2. $\frac{3}{5} \times \frac{1}{4}$
3. $\frac{7}{8} \times \frac{1}{6}$
4. $\frac{7}{9} \times \frac{2}{3}$
5. $\frac{4}{9} \times \frac{1}{7}$
6. $\frac{6}{11} \times \frac{5}{8}$
7. $\frac{4}{7} \times \frac{1}{3}$
8. $\frac{1}{2} \times \frac{3}{7}$
9. $\frac{7}{11} \times \frac{1}{5}$
10. $\frac{1}{4} \times \frac{3}{4}$
11. $8 \times \frac{3}{4}$
12. $\frac{2}{9} \times \frac{1}{6}$
13. $\frac{4}{7} \times \frac{1}{8}$
14. $\frac{11}{12} \times \frac{2}{5}$
15. $\frac{3}{8} \times \frac{4}{5}$
16. $\frac{7}{15} \times \frac{5}{8}$
17. $\frac{3}{5} \times \frac{5}{9}$
18. $\frac{7}{16} \times \frac{8}{9}$
19. $\frac{3}{7} \times \frac{2}{9}$
20. $\frac{6}{7} \times \frac{1}{2}$

Practice Problems

1. $\frac{3}{5} \times \frac{1}{2}$ 2. $\frac{6}{7} \times \frac{2}{3}$ 3. $\frac{3}{8} \times \frac{1}{4}$ 4. $\frac{3}{7} \times \frac{8}{9}$

5. $\frac{1}{2} \times \frac{3}{10}$ 6. $\frac{7}{9} \times \frac{3}{5}$ 7. $\frac{6}{11} \times 3$ 8. $\frac{4}{5} \times \frac{1}{8}$

9. $\frac{1}{61} \times \frac{3}{5}$ 10. $\frac{7}{9} \times \frac{1}{14}$ 11. $8 \times \frac{3}{4}$ 12. $\frac{2}{9} \times \frac{1}{6}$

13. $\frac{16}{21} \times \frac{7}{8}$ 14. $\frac{9}{10} \times \frac{2}{3}$ 15. $\frac{6}{7} \times \frac{1}{4}$ 16. $9 \times \frac{2}{5}$

17. $\frac{14}{17} \times \frac{8}{9}$ 18. $\frac{3}{8} \times \frac{2}{5}$ 19. $\frac{4}{9} \times \frac{3}{7}$ 20. $\frac{3}{5} \times \frac{4}{11}$

21. $\frac{6}{7} \times \frac{1}{3}$ 22. $\frac{9}{13} \times \frac{4}{5}$ 23. $\frac{3}{8} \times 6$ 24. $\frac{5}{6} \times 8$

25. $\frac{4}{15} \times \frac{5}{6}$ 26. $\frac{1}{2} \times \frac{2}{3}$ 27. $\frac{8}{9} \times \frac{9}{11}$ 28. $\frac{3}{4} \times 7$

29. $\frac{6}{17} \times \frac{4}{5}$ 30. $\frac{2}{9} \times \frac{3}{8}$ 31. $\frac{7}{16} \times \frac{4}{5}$ 32. $\frac{3}{10} \times \frac{5}{8}$

33. $\frac{2}{3} \times \frac{7}{9}$ 34. $\frac{3}{14} \times \frac{7}{12}$ 35. $\frac{4}{5} \times \frac{1}{6}$ 36. $\frac{8}{11} \times \frac{3}{4}$

37. $\frac{1}{12} \times \frac{2}{5}$ 38. $\frac{16}{17} \times \frac{3}{4}$ 39. $\frac{1}{9} \times \frac{5}{18}$ 40. $\frac{7}{9} \times \frac{1}{3}$

41. $\frac{11}{14} \times \frac{7}{9}$ 42. $\frac{5}{8} \times \frac{3}{10}$ 43. $\frac{1}{3} \times \frac{4}{5}$ 44. $\frac{6}{7} \times \frac{3}{14}$

45. $\frac{5}{6} \times \frac{3}{5}$ 46. $\frac{12}{17} \times \frac{34}{35}$ 47. $\frac{1}{9} \times \frac{3}{8}$ 48. $\frac{4}{7} \times \frac{3}{5}$

Now you're ready to multiply a fraction and a mixed number. It's simple! First, you must change the mixed number to an improper fraction, then multiply as you did earlier.

$$\frac{2}{5} \times 2\frac{1}{4} = \frac{2}{5} \times \frac{9}{4} = \frac{18}{20} = \frac{9}{10}$$

One more example

$$5\frac{2}{3} \times \frac{6}{7} = \frac{17}{3} \times \frac{6}{7}$$

Change the mixed number to an improper fraction.

$$= \frac{17 \times 6}{3 \times 7}$$

Multiply numerators, then multiply denominators.

$$= \frac{102}{21}$$

$$= 4\frac{18}{21} = 4\frac{6}{7}$$

Change the improper fraction in the answer to a mixed number and reduce.

Now it's your turn. Try these:

1. $\frac{3}{4} \times 3\frac{1}{2}$

2. $\frac{5}{8} \times 6\frac{1}{3}$

3. $7\frac{1}{9} \times \frac{3}{5}$

4. $2\frac{2}{3} \times \frac{1}{2}$

Answers

1. $\frac{3}{4} \times 3\frac{1}{2} = \frac{3}{4} \times \frac{7}{2} = \frac{3 \times 7}{4 \times 2} = \frac{21}{8} = 2\frac{5}{8}$

2. $\frac{5}{8} \times 6\frac{1}{3} = \frac{5}{8} \times \frac{19}{3} = \frac{5 \times 19}{8 \times 3} = \frac{95}{24} = 3\frac{23}{24}$

3. $7\frac{1}{9} \times \frac{3}{5} = \frac{64}{9} \times \frac{3}{5} = \frac{64 \times 3}{9 \times 5} = \frac{192}{45} = 4\frac{12}{45} = 4\frac{4}{15}$

4. $2\frac{2}{3} \times \frac{1}{2} = \frac{8}{3} \times \frac{1}{2} = \frac{8 \times 1}{3 \times 2} = \frac{8}{6} = 1\frac{2}{6} = 1\frac{1}{3}$

If you had trouble, go back to page 49.

Here is a way to help you make some problems a little easier. In problems 3 and 4, you could have reduced before you multiplied. This will keep you from multiplying such big numbers and reducing your answers.

In problem 3, you could have reduced the 3 and 9, like this: $\frac{64 \times \cancel{3}^{1}}{{}_{3}\cancel{9} \times 5} = \frac{64 \times 1}{3 \times 5} = \frac{64}{15} = 4\frac{4}{5}$

In problem 4, you could have reduced the 8 and 2, like this: $\frac{{}^{4}\cancel{8} \times 1}{3 \times \cancel{2}_{1}} = \frac{4 \times 1}{3 \times 3} = \frac{4}{3} = 1\frac{1}{3}$

Practice Problems

1. $\frac{3}{5} \times 1\frac{1}{2}$	2. $\frac{5}{8} \times 2\frac{1}{4}$	3. $6\frac{2}{5} \times \frac{5}{8}$	4. $3\frac{1}{2} \times \frac{3}{5}$
5. $4\frac{2}{5} \times \frac{1}{7}$	6. $7\frac{2}{3} \times \frac{1}{6}$	7. $\frac{4}{9} \times 1\frac{1}{4}$	8. $5\frac{1}{8} \times \frac{2}{3}$
9. $\frac{3}{8} \times 2\frac{3}{4}$	10. $3\frac{1}{6} \times \frac{3}{5}$	11. $8\frac{1}{2} \times \frac{7}{9}$	12. $\frac{5}{6} \times 2\frac{1}{3}$
13. $\frac{1}{5} \times 3\frac{1}{2}$	14. $\frac{6}{7} \times 2\frac{1}{6}$	15. $\frac{7}{8} \times 4\frac{1}{3}$	16. $\frac{1}{10} \times 6\frac{1}{4}$
17. $5\frac{1}{9} \times \frac{3}{4}$	18. $\frac{5}{7} \times 2\frac{1}{2}$	19. $\frac{7}{9} \times 4\frac{1}{5}$	20. $\frac{4}{5} \times 3\frac{1}{4}$
21. $\frac{11}{12} \times 1\frac{2}{3}$	22. $5\frac{4}{9} \times \frac{2}{5}$	23. $\frac{3}{5} \times 2\frac{1}{3}$	24. $\frac{6}{7} \times 4\frac{3}{4}$

Practice Problems

1. $\frac{1}{4} \times 3\frac{2}{5}$	2. $\frac{5}{8} \times 2\frac{1}{3}$	3. $\frac{4}{7} \times 6\frac{1}{2}$	4. $5\frac{2}{3} \times \frac{5}{6}$
5. $\frac{3}{4} \times 1\frac{7}{8}$	6. $4\frac{1}{3} \times \frac{3}{4}$	7. $\frac{1}{9} \times 2\frac{5}{8}$	8. $\frac{1}{7} \times 6\frac{2}{3}$
9. $\frac{4}{9} \times 7\frac{1}{2}$	10. $9\frac{1}{10} \times \frac{5}{9}$	11. $\frac{6}{11} \times 2\frac{1}{8}$	12. $\frac{3}{4} \times 6\frac{1}{5}$
13. $4\frac{2}{5} \times \frac{1}{6}$	14. $1\frac{1}{3} \times \frac{3}{7}$	15. $\frac{1}{8} \times 3\frac{3}{4}$	16. $\frac{5}{6} \times 1\frac{1}{9}$
17. $\frac{6}{7} \times 3\frac{1}{2}$	18. $5\frac{1}{3} \times \frac{2}{5}$	19. $\frac{7}{9} \times 3\frac{1}{4}$	20. $2\frac{5}{7} \times \frac{7}{8}$
21. $7\frac{1}{2} \times \frac{3}{5}$	22. $8\frac{1}{4} \times \frac{4}{5}$	23. $\frac{1}{5} \times 2\frac{2}{3}$	24. $\frac{1}{10} \times 3\frac{5}{8}$
25. $4\frac{2}{5} \times \frac{1}{6}$	26. $1\frac{1}{9} \times \frac{3}{4}$	27. $5\frac{3}{4} \times \frac{4}{7}$	28. $6\frac{6}{7} \times \frac{3}{5}$
29. $4\frac{1}{5} \times \frac{3}{4}$	30. $\frac{6}{7} \times 1\frac{9}{10}$	31. $\frac{7}{8} \times 4\frac{3}{5}$	32. $\frac{1}{12} \times 3\frac{4}{5}$
33. $\frac{4}{7} \times 2\frac{1}{4}$	34. $8\frac{1}{9} \times \frac{3}{4}$	35. $\frac{7}{10} \times 2\frac{1}{3}$	36. $\frac{4}{5} \times 5\frac{3}{8}$
37. $\frac{6}{7} \times 3\frac{1}{2}$	38. $\frac{9}{10} \times 2\frac{2}{3}$	39. $7\frac{1}{2} \times \frac{5}{6}$	40. $8\frac{1}{3} \times \frac{6}{7}$
41. $\frac{3}{8} \times 2\frac{1}{6}$	42. $\frac{3}{10} \times 4\frac{1}{5}$	43. $\frac{1}{12} \times 2\frac{1}{8}$	44. $\frac{1}{4} \times 5\frac{5}{6}$
45. $\frac{3}{4} \times 7\frac{3}{8}$	46. $3\frac{2}{5} \times \frac{5}{8}$	47. $\frac{5}{6} \times 3\frac{3}{5}$	48. $\frac{7}{10} \times 2\frac{5}{8}$

To multiply two mixed numbers, change both of them to an improper fraction and multiply as before.

$$3\frac{1}{3} \times 2\frac{2}{5} = \frac{10}{3} \times \frac{12}{5}$$

Change both to improper fractions.

$$= \frac{10 \times 12}{3 \times 5}$$

Multiply numerators and then multiply denominators.

$$= \frac{10 \times \cancel{12}^{\,4}}{{}_{1}\cancel{3} \times 5}$$

Reduce now if you want to.

$$= \frac{\cancel{10}^{\,2} \times 4}{1 \times \cancel{5}_{\,1}}$$

Reduce even more if you can.

$$= \frac{2 \times 4}{1 \times 1}$$

Now you have easy numbers to multiply.

$$= \frac{8}{1}$$

$$= 8$$

You do these:

1. $6\frac{1}{4} \times 2\frac{3}{7}$

2. $9\frac{3}{5} \times 4\frac{1}{4}$

3. $5\frac{1}{3} \times 3\frac{3}{8}$

4. $4\frac{2}{7} \times 3\frac{4}{5}$

Answers

1. $6\frac{1}{4} \times 2\frac{3}{7} = \frac{25}{4} \times \frac{17}{7} = \frac{425}{28} = 15\frac{5}{28}$

2. $9\frac{3}{5} \times 4\frac{1}{4} = \frac{\overset{12}{\cancel{48}}}{5} \times \frac{17}{\underset{1}{\cancel{4}}} = \frac{204}{5} = 40\frac{4}{5}$

3. $5\frac{1}{3} \times 3\frac{3}{8} = \frac{\overset{2}{\cancel{16}}}{\underset{1}{\cancel{3}}} \times \frac{\overset{9}{\cancel{27}}}{\underset{1}{\cancel{8}}} = \frac{18}{1} = 18$

4. $4\frac{2}{7} \times 3\frac{4}{5} = \frac{\overset{6}{\cancel{30}}}{7} \times \frac{19}{\underset{1}{\cancel{5}}} = \frac{114}{7} = 16\frac{2}{7}$

If you missed one, it was probably a multiplication or division mistake.

Notice that in problem 3, reducing early made this problem much easier.

Be careful with your multiplying and dividing and you'll have no trouble.

Practice Problems

1. $2\frac{1}{3} \times 3\frac{2}{3}$	2. $4\frac{1}{2} \times 1\frac{3}{4}$	3. $6\frac{1}{4} \times 2\frac{1}{8}$	4. $1\frac{1}{9} \times 2\frac{2}{5}$
5. $4\frac{1}{8} \times 2\frac{2}{5}$	6. $7\frac{1}{2} \times 2\frac{2}{3}$	7. $8\frac{1}{2} \times 2\frac{3}{4}$	8. $6\frac{1}{3} \times 2\frac{3}{4}$
9. $5\frac{3}{8} \times 3\frac{1}{3}$	10. $7\frac{2}{5} \times 4\frac{1}{2}$	11. $6\frac{3}{4} \times 1\frac{4}{5}$	12. $3\frac{3}{8} \times 2\frac{1}{4}$
13. $2\frac{1}{6} \times 4\frac{2}{3}$	14. $5\frac{1}{7} \times 1\frac{1}{8}$	15. $1\frac{7}{8} \times 4\frac{1}{6}$	16. $9\frac{1}{3} \times 3\frac{3}{4}$

Practice Problems

1. $4\frac{1}{2} \times 1\frac{1}{3}$	2. $6\frac{2}{3} \times 2\frac{1}{6}$	3. $5\frac{1}{4} \times 3\frac{2}{9}$	4. $2\frac{1}{10} \times 3\frac{2}{3}$
5. $7\frac{1}{6} \times 2\frac{3}{4}$	6. $4\frac{1}{5} \times 6\frac{1}{2}$	7. $8\frac{1}{4} \times 1\frac{1}{6}$	8. $3\frac{2}{5} \times 1\frac{1}{7}$
9. $4\frac{1}{9} \times 2\frac{2}{3}$	10. $6\frac{3}{4} \times 1\frac{1}{8}$	11. $9\frac{1}{3} \times 2\frac{7}{8}$	12. $3\frac{5}{6} \times 2\frac{1}{4}$
13. $5\frac{4}{9} \times 2\frac{1}{6}$	14. $8\frac{1}{3} \times 1\frac{1}{2}$	15. $7\frac{1}{8} \times 4\frac{1}{6}$	16. $4\frac{5}{6} \times 2\frac{1}{2}$
17. $7\frac{1}{8} \times 1\frac{3}{8}$	18. $5\frac{1}{2} \times 3\frac{3}{5}$	19. $4\frac{5}{8} \times 3\frac{1}{7}$	20. $6\frac{5}{9} \times 1\frac{1}{2}$
21. $5\frac{1}{3} \times 2\frac{2}{3}$	22. $7\frac{1}{4} \times 3\frac{2}{3}$	23. $5\frac{1}{6} \times 4\frac{2}{5}$	24. $6\frac{1}{2} \times 1\frac{5}{8}$
25. $1\frac{1}{6} \times 4\frac{2}{3}$	26. $8\frac{1}{2} \times 2\frac{2}{5}$	27. $6\frac{1}{4} \times 3\frac{1}{6}$	28. $8\frac{1}{3} \times 1\frac{1}{7}$
29. $4\frac{1}{2} \times 1\frac{5}{7}$	30. $3\frac{9}{10} \times 2\frac{1}{5}$	31. $5\frac{2}{3} \times 2\frac{1}{4}$	32. $7\frac{3}{4} \times 2\frac{1}{8}$
33. $5\frac{1}{4} \times 6\frac{1}{2}$	34. $7\frac{2}{3} \times 1\frac{1}{8}$	35. $1\frac{1}{7} \times 2\frac{2}{3}$	36. $9\frac{1}{2} \times 3\frac{3}{4}$
37. $8\frac{1}{4} \times 3\frac{2}{5}$	38. $6\frac{6}{7} \times 1\frac{1}{5}$	39. $2\frac{1}{6} \times 3\frac{1}{8}$	40. $6\frac{1}{9} \times 3\frac{2}{3}$
41. $4\frac{5}{6} \times 3\frac{1}{8}$	42. $7\frac{6}{11} \times 1\frac{1}{2}$	43. $9\frac{1}{10} \times 2\frac{2}{5}$	44. $6\frac{1}{3} \times 2\frac{2}{9}$
45. $5\frac{1}{5} \times 3\frac{3}{4}$	46. $7\frac{1}{2} \times 2\frac{2}{3}$	47. $4\frac{3}{4} \times 1\frac{1}{8}$	48. $1\frac{7}{8} \times 4\frac{1}{2}$
49. $6\frac{2}{3} \times 3\frac{1}{6}$	50. $1\frac{1}{5} \times 2\frac{3}{5}$	51. $6\frac{3}{5} \times 4\frac{1}{4}$	52. $9\frac{1}{2} \times 2\frac{3}{10}$
53. $3\frac{2}{3} \times 1\frac{1}{4}$	54. $4\frac{1}{6} \times 3\frac{1}{2}$	55. $4\frac{3}{5} \times 2\frac{1}{8}$	56. $5\frac{1}{7} \times 2\frac{1}{2}$

Practice Problems

Reduce answers:

1. $\frac{1}{2} \times \frac{2}{3}$	2. $\frac{2}{7} \times \frac{4}{7}$	3. $\frac{3}{15} \times \frac{2}{5}$	4. $\frac{6}{11} \times \frac{1}{3}$
5. $\frac{5}{6} \times \frac{3}{4}$	6. $\frac{3}{7} \times \frac{9}{12}$	7. $\frac{7}{9} \times \frac{1}{14}$	8. $\frac{6}{7} \times \frac{2}{3}$
9. $\frac{5}{8} \times \frac{1}{4}$	10. $\frac{12}{17} \times \frac{5}{9}$	11. $\frac{6}{13} \times \frac{1}{12}$	12. $\frac{4}{9} \times \frac{3}{8}$
13. $8 \times \frac{1}{2}$	14. $12 \times \frac{2}{3}$	15. $4 \times \frac{1}{5}$	16. $9 \times \frac{1}{9}$
17. $\frac{4}{5} \times 20$	18. $\frac{7}{8} \times 3$	19. $\frac{3}{7} \times 12$	20. $\frac{4}{9} \times 20$
21. $\frac{3}{5} \times 1\frac{2}{3}$	22. $\frac{4}{7} \times 3\frac{1}{2}$	23. $\frac{1}{18} \times 6\frac{1}{4}$	24. $\frac{5}{8} \times 16\frac{1}{8}$
25. $4\frac{1}{2} \times \frac{2}{5}$	26. $6\frac{3}{5} \times \frac{1}{9}$	27. $5\frac{5}{9} \times \frac{1}{3}$	28. $12\frac{1}{2} \times \frac{2}{3}$
29. $2\frac{1}{2} \times 1\frac{4}{5}$	30. $4\frac{2}{3} \times 3\frac{1}{5}$	31. $5\frac{1}{8} \times 3\frac{1}{5}$	32. $7\frac{1}{7} \times 2\frac{3}{5}$
33. $3\frac{9}{10} \times 2\frac{1}{6}$	34. $6\frac{3}{8} \times 2\frac{1}{4}$	35. $4\frac{11}{12} \times 2\frac{3}{4}$	36. $9\frac{1}{8} \times 2\frac{1}{9}$
37. $4\frac{7}{10} \times 1\frac{1}{5}$	38. $9\frac{1}{9} \times 2\frac{4}{11}$	39. $12\frac{1}{4} \times 8\frac{3}{8}$	40. $19\frac{1}{3} \times 6\frac{1}{4}$

Practice Problems

Reduce answers:

1. $\frac{2}{9} \times \frac{1}{4}$ 2. $\frac{6}{11} \times \frac{1}{2}$ 3. $\frac{3}{4} \times \frac{1}{4}$ 4. $\frac{7}{9} \times \frac{1}{5}$

5. $\frac{7}{12} \times \frac{3}{4}$ 6. $\frac{3}{5} \times \frac{7}{13}$ 7. $\frac{8}{15} \times \frac{3}{8}$ 8. $\frac{6}{7} \times \frac{7}{9}$

9. $\frac{14}{15} \times \frac{5}{7}$ 10. $\frac{4}{5} \times \frac{10}{17}$ 11. $\frac{8}{9} \times \frac{7}{10}$ 12. $\frac{3}{4} \times \frac{9}{11}$

13. $4 \times \frac{2}{3}$ 14. $9 \times \frac{1}{6}$ 15. $16 \times \frac{2}{7}$ 16. $20 \times \frac{2}{5}$

17. $\frac{3}{11} \times 4$ 18. $\frac{7}{8} \times 12$ 19. $\frac{5}{6} \times 3$ 20. $\frac{4}{9} \times 30$

21. $\frac{3}{4} \times 1\frac{1}{5}$ 22. $\frac{1}{2} \times 4\frac{1}{2}$ 23. $\frac{6}{7} \times 3\frac{1}{3}$ 24. $\frac{5}{8} \times 6\frac{1}{4}$

25. $4\frac{1}{4} \times \frac{2}{3}$ 26. $5\frac{1}{9} \times \frac{1}{8}$ 27. $1\frac{2}{7} \times \frac{7}{9}$ 28. $16\frac{1}{2} \times \frac{3}{11}$

29. $5\frac{1}{4} \times 1\frac{1}{3}$ 30. $6\frac{2}{3} \times 2\frac{1}{8}$ 31. $4\frac{4}{7} \times 3\frac{1}{16}$ 32. $2\frac{2}{7} \times 3\frac{1}{3}$

33. $8\frac{1}{2} \times 6\frac{3}{4}$ 34. $9\frac{1}{5} \times 1\frac{2}{23}$ 35. $10\frac{1}{4} \times 2\frac{3}{8}$ 36. $3\frac{1}{5} \times 2\frac{2}{7}$

37. $7\frac{2}{3} \times 1\frac{1}{6}$ 38. $4\frac{1}{6} \times 2\frac{2}{5}$ 39. $7\frac{9}{10} \times 2\frac{1}{2}$ 40. $2\frac{3}{7} \times 3\frac{3}{17}$

Using Multiplication of Fractions

1. Jim works $9\frac{1}{2}$ hours each day, Monday through Friday. How many hours did he work?

2. Bill has 55 matches that are $1\frac{3}{4}$ inches long and he places them end to end. How long is the line of matches?

3. Hannah bought 14 pieces of material on sale. Each piece of material was $2\frac{3}{4}$ yards. How many yards of material did she buy?

4. Sebastian jogs $3\frac{1}{2}$ miles each day of the week. How many miles does he jog in a week?

5. Carla usually walks $1\frac{3}{4}$ miles in one lap. The route she takes is three laps around her neighborhood a day. How many miles does she walk in 2 days?

Using Multiplication of Fractions (cont.)

6. Jeremy's mom left $\frac{1}{2}$ of the cake for him and his two friends to divide equally. If Jeremy and his friends get $\frac{1}{3}$ of the half of cake, how much of the whole cake did each person get?

7. Jackie buys $6\frac{1}{2}$ pounds of peanuts and gives each of her four children and herself equal amounts of the peanuts. Each person then receives $\frac{1}{5}$ of the peanuts. How many pounds of peanuts does each of them get?

8. If five boards, measuring $3\frac{3}{8}$ feet long each, were laid end to end, what would be the total length.

9. A recipe calls for $\frac{2}{3}$ stick of butter. How much butter would you need if you double the recipe?

10. The American Textile Company has 1800 employees. If four-fifth of their employees belong to the union, how many employees do not belong to the union?

Dividing Fractions

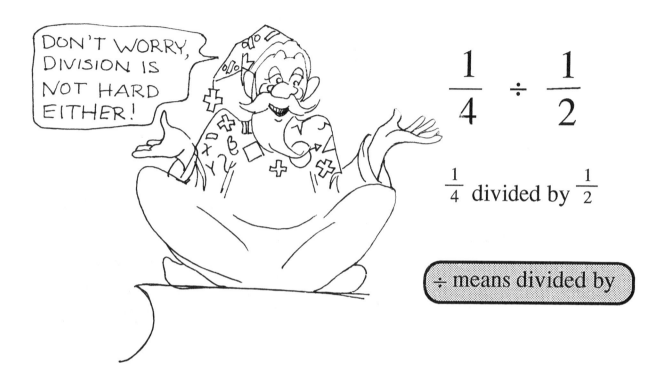

$$\frac{1}{4} \div \frac{1}{2}$$

$\frac{1}{4}$ divided by $\frac{1}{2}$

$\div$ means divided by

Following this lesson, you will be able to:

- Divide using two common fractions.
- Divide using a common fraction and a whole number.
- Divide using a common fraction and a mixed number.
- Divide using two mixed numbers.

If you can multiply fractions, you will be able to divide fractions with no problem!

When you divide with fractions, just invert the number that is "divided by" and then multiply as usual.

$$\frac{1}{4} \div \frac{1}{2}$$

$$\frac{1}{4} \times \frac{2}{1}$$

Invert $\frac{1}{2}$ to $\frac{2}{1}$

$$\frac{1 \times 2}{4 \times 1}$$ (Now, just multiply.)

Invert means to flip the fraction around so the numerator becomes the denominator and the denominator becomes the numerator.

$$\frac{2}{4}$$ (Then reduce.)

$$\frac{1}{2}$$

You do the same if one of the numbers is a whole number.

Just remember that 5 is $\frac{5}{1}$ and $\frac{5}{1}$ inverted is $\frac{1}{5}$.

Example: $\frac{2}{3} \div 5 = \frac{2}{3} \times \frac{1}{5} = \frac{2}{15}$

You try these:

1. $\frac{3}{4} \div \frac{1}{4}$

2. $\frac{7}{10} \div \frac{3}{5}$

3. $6 \div \frac{2}{7}$

4. $\frac{4}{9} \div 8$

Answers

1. $\frac{3}{4} \div \frac{1}{4} = \frac{3}{4} \times \frac{4}{1} = \frac{12}{4} = 3$

2. $\frac{7}{10} \div \frac{3}{5} = \frac{7}{10} \times \frac{5}{3} = \frac{35}{30} = 1\frac{5}{30} = 1\frac{1}{6}$

3. $6 \div \frac{2}{7} = \frac{6}{1} \times \frac{7}{2} = \frac{42}{2} = 21$

4. $\frac{4}{9} \div 8 = \frac{4}{9} \times \frac{1}{8} = \frac{4}{72} = \frac{1}{18}$

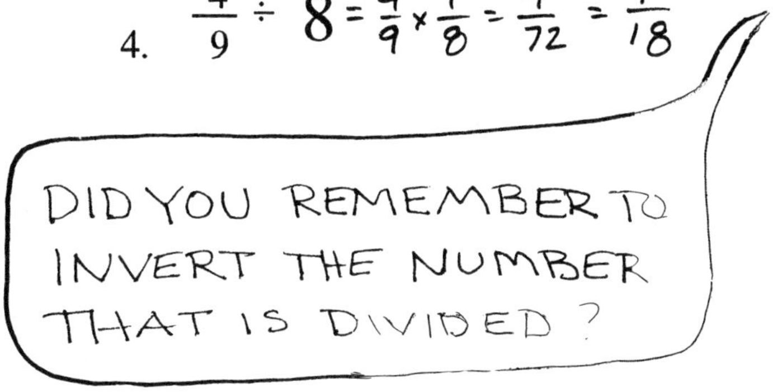
DID YOU REMEMBER TO INVERT THE NUMBER THAT IS DIVIDED ?

The examples in the rest of this lesson will not be reduced before multiplying so that you can see the step. But it is easier for you if you do reduce before multiplying.

Practice Problems

1. $\frac{1}{3} \div \frac{2}{5}$	2. $\frac{3}{8} \div \frac{1}{4}$	3. $\frac{7}{9} \div \frac{1}{3}$	4. $\frac{2}{9} \div \frac{3}{4}$	5. $\frac{1}{2} \div \frac{1}{4}$
6. $\frac{6}{10} \div \frac{1}{5}$	7. $\frac{7}{8} \div \frac{2}{3}$	8. $\frac{6}{7} \div \frac{3}{4}$	9. $\frac{5}{9} \div \frac{1}{4}$	10. $\frac{7}{10} \div \frac{3}{4}$
11. $\frac{6}{11} \div \frac{1}{5}$	12. $\frac{3}{5} \div \frac{1}{6}$	13. $\frac{5}{9} \div \frac{2}{5}$	14. $\frac{5}{7} \div \frac{3}{4}$	15. $\frac{5}{8} \div \frac{4}{9}$
16. $\frac{1}{2} \div \frac{6}{13}$	17. $\frac{7}{15} \div \frac{2}{3}$	18. $\frac{7}{20} \div \frac{1}{6}$	19. $\frac{4}{5} \div \frac{2}{3}$	20. $\frac{8}{19} \div \frac{1}{5}$
21. $\frac{2}{13} \div \frac{4}{5}$	22. $\frac{7}{8} \div \frac{1}{3}$	23. $\frac{7}{9} \div \frac{4}{5}$	24. $\frac{6}{17} \div \frac{1}{3}$	25. $\frac{2}{9} \div \frac{1}{9}$

Practice Problems

1. $\frac{6}{7} \div \frac{2}{5}$	2. $\frac{3}{4} \div \frac{1}{6}$	3. $\frac{7}{8} \div \frac{3}{8}$	4. $\frac{2}{5} \div \frac{7}{10}$	5. $\frac{9}{11} \div \frac{3}{4}$
6. $8 \div \frac{3}{5}$	7. $\frac{2}{9} \div \frac{1}{4}$	8. $\frac{2}{5} \div \frac{1}{7}$	9. $\frac{3}{8} \div \frac{7}{10}$	10. $\frac{4}{5} \div \frac{1}{2}$
11. $\frac{9}{11} \div 6$	12. $\frac{3}{16} \div \frac{2}{3}$	13. $\frac{5}{8} \div \frac{1}{17}$	14. $\frac{3}{10} \div 7$	15. $6 \div \frac{7}{8}$
16. $\frac{3}{4} \div \frac{2}{9}$	17. $\frac{6}{13} \div 4$	18. $\frac{3}{5} \div \frac{5}{6}$	19. $\frac{1}{2} \div \frac{3}{17}$	20. $\frac{2}{9} \div \frac{7}{15}$
21. $\frac{12}{13} \div \frac{1}{4}$	22. $\frac{5}{8} \div \frac{1}{3}$	23. $\frac{7}{9} \div \frac{4}{5}$	24. $\frac{7}{22} \div \frac{1}{2}$	25. $\frac{5}{6} \div \frac{1}{8}$
26. $\frac{7}{8} \div \frac{1}{4}$	27. $\frac{1}{9} \div \frac{7}{10}$	28. $\frac{4}{5} \div \frac{1}{6}$	29. $\frac{5}{8} \div \frac{7}{10}$	30. $\frac{4}{7} \div 8$
31. $9 \div \frac{3}{5}$	32. $\frac{1}{4} \div \frac{1}{3}$	33. $\frac{1}{6} \div 9$	34. $\frac{2}{5} \div \frac{1}{7}$	35. $\frac{3}{10} \div \frac{6}{7}$
36. $\frac{5}{9} \div \frac{3}{4}$	37. $\frac{2}{19} \div \frac{1}{8}$	38. $\frac{21}{25} \div \frac{1}{5}$	39. $\frac{6}{13} \div 4$	40. $\frac{2}{9} \div \frac{1}{18}$
41. $\frac{7}{8} \div 4$	42. $\frac{3}{19} \div \frac{1}{6}$	43. $12 \div \frac{2}{3}$	44. $\frac{6}{11} \div 4$	45. $\frac{1}{2} \div \frac{4}{5}$
46. $\frac{9}{11} \div \frac{3}{8}$	47. $\frac{6}{23} \div \frac{3}{5}$	48. $\frac{7}{10} \div \frac{1}{7}$	49. $6 \div \frac{1}{8}$	50. $\frac{4}{5} \div \frac{1}{8}$

Now, you can divide a fraction and a mixed number. Change the mixed number to an improper fraction first, then invert and multiply.

Here's how you divide a mixed number by a proper fraction:

$$2\frac{1}{3} \div \frac{1}{4}$$

$$= \frac{7}{3} \div \frac{1}{4}$$

Change the mixed number to an improper fraction.

$$= \frac{7}{3} \times \frac{4}{1}$$

Invert the number divided by and multiply.

$$= \frac{28}{3}$$

$$= 9\frac{1}{3}$$

In the answer, change the improper fraction to a mixed number.

Look at this example where the mixed number is the one "divided by."
There is a mistake made in this problem. Can you tell where the mistake is made?

$$\frac{5}{6} \div 3\frac{1}{3}$$

$$= \frac{5}{6} \times \frac{10}{3}$$

$$= \frac{50}{18}$$

$$= 2\frac{14}{18}$$

$$= 2\frac{7}{9}$$

Did you find the mistake?

Look: This is correct.

$$\frac{5}{6} \div 3\frac{1}{3}$$

$$= \frac{5}{6} \div \frac{10}{3}$$

The mistake was in doing too much at one time. The mixed number was changed to an improper fraction and division was changed to multiplication in one step. Watch out and don't make this common mistake!

$$= \frac{5}{6} \times \frac{3}{10}$$

$$= \frac{15}{60}$$

$$= \frac{1}{4}$$

You try these:

1. $\frac{1}{5} \div 2\frac{1}{4}$

2. $4\frac{2}{3} \div \frac{1}{8}$

3. $4\frac{1}{9} \div 6$

Answers

1. $\frac{1}{5} \div 2\frac{1}{4} = \frac{1}{5} \div \frac{9}{4} = \frac{1}{5} \times \frac{4}{9} = \frac{4}{45}$

2. $4\frac{2}{3} \div \frac{1}{8} = \frac{14}{3} \div \frac{1}{8} = \frac{14}{3} \times \frac{8}{1} = \frac{112}{3} = 37\frac{1}{3}$

3. $4\frac{1}{9} \div 6 = \frac{37}{9} \div \frac{6}{1} = \frac{37}{9} \times \frac{1}{6} = \frac{37}{54}$

Did you change the mixed number to an improper fraction before you inverted?

Notice that in problem 3, one of the numbers was a whole number. **As a reminder:** *Whole numbers have a denominator of 1 and can be treated like any other improper fraction.*

Practice Problems

1. $2\frac{1}{4} \div \frac{1}{3}$ 2. $3\frac{1}{2} \div \frac{1}{6}$ 3. $\frac{1}{2} \div 1\frac{1}{4}$ 4. $7\frac{2}{3} \div \frac{4}{5}$ 5. $6\frac{1}{2} \div \frac{3}{4}$

6. $2\frac{1}{9} \div \frac{1}{4}$ 7. $6\frac{2}{3} \div \frac{4}{5}$ 8. $9\frac{1}{3} \div \frac{2}{3}$ 9. $6\frac{2}{3} \div \frac{1}{4}$ 10. $\frac{7}{8} \div 2\frac{1}{4}$

11. $\frac{3}{4} \div 3\frac{1}{3}$ 12. $6\frac{1}{5} \div \frac{2}{3}$ 13. $\frac{9}{10} \div 1\frac{1}{4}$ 14. $\frac{7}{11} \div 2\frac{1}{6}$ 15. $6\frac{1}{8} \div \frac{7}{9}$

16. $\frac{4}{9} \div 2\frac{1}{2}$ 17. $4\frac{3}{4} \div \frac{3}{4}$ 18. $6\frac{1}{6} \div \frac{6}{11}$ 19. $5\frac{1}{4} \div \frac{3}{5}$ 20. $\frac{7}{9} \div 2\frac{1}{3}$

21. $6\frac{1}{8} \div \frac{7}{8}$ 22. $5\frac{1}{9} \div \frac{7}{8}$ 23. $4\frac{1}{3} \div \frac{1}{2}$ 24. $6\frac{5}{9} \div \frac{3}{4}$ 25. $\frac{12}{13} \div 1\frac{1}{4}$

26. $\frac{3}{5} \div 2\frac{1}{3}$ 27. $4\frac{1}{8} \div \frac{4}{7}$ 28. $2\frac{1}{6} \div \frac{5}{7}$ 29. $3\frac{1}{3} \div \frac{2}{9}$ 30. $7\frac{9}{10} \div \frac{4}{5}$

Practice Problems

1. $7\frac{1}{4} \div \frac{4}{5}$ 2. $3\frac{1}{2} \div \frac{1}{9}$ 3. $\frac{7}{9} \div 2\frac{1}{4}$ 4. $\frac{6}{7} \div 1\frac{1}{3}$ 5. $4\frac{2}{3} \div \frac{3}{5}$

6. $7\frac{2}{5} \div \frac{7}{10}$ 7. $4\frac{1}{6} \div \frac{2}{3}$ 8. $2\frac{1}{4} \div \frac{7}{8}$ 9. $6\frac{1}{4} \div \frac{3}{4}$ 10. $\frac{9}{10} \div 3\frac{1}{5}$

11. $8\frac{1}{2} \div \frac{4}{5}$ 12. $\frac{7}{11} \div 2\frac{2}{3}$ 13. $\frac{11}{12} \div 3\frac{1}{4}$ 14. $\frac{6}{7} \div 1\frac{1}{5}$ 15. $4\frac{1}{9} \div \frac{7}{9}$

16. $\frac{7}{12} \div 1\frac{1}{5}$ 17. $7\frac{1}{6} \div \frac{3}{4}$ 18. $9\frac{1}{2} \div \frac{7}{9}$ 19. $\frac{13}{14} \div 1\frac{3}{5}$ 20. $6\frac{3}{8} \div \frac{4}{5}$

21. $3\frac{1}{2} \div \frac{2}{3}$ 22. $6\frac{1}{6} \div \frac{7}{10}$ 23. $6\frac{1}{5} \div \frac{2}{3}$ 24. $\frac{7}{8} \div 2\frac{1}{4}$ 25. $8\frac{1}{8} \div \frac{3}{5}$

26. $7\frac{1}{2} \div \frac{3}{4}$ 27. $\frac{9}{13} \div 2\frac{1}{3}$ 28. $4\frac{1}{3} \div \frac{2}{3}$ 29. $4\frac{2}{5} \div \frac{4}{7}$ 30. $\frac{11}{12} \div 2\frac{1}{5}$

31. $4\frac{1}{4} \div \frac{2}{3}$ 32. $7\frac{1}{9} \div \frac{3}{4}$ 33. $5\frac{1}{6} \div \frac{5}{7}$ 34. $\frac{7}{8} \div 1\frac{1}{2}$ 35. $6\frac{1}{2} \div \frac{7}{8}$

36. $\frac{3}{10} \div 2\frac{1}{4}$ 37. $4\frac{1}{5} \div \frac{1}{6}$ 38. $\frac{9}{17} \div 2\frac{1}{3}$ 39. $\frac{7}{9} \div 1\frac{1}{3}$ 40. $3\frac{1}{8} \div \frac{1}{4}$

41. $2\frac{1}{2} \div \frac{3}{4}$ 42. $6\frac{2}{5} \div \frac{1}{5}$ 43. $7\frac{1}{2} \div \frac{9}{10}$ 44. $5\frac{1}{9} \div \frac{3}{8}$ 45. $6\frac{2}{9} \div \frac{1}{4}$

Dividing two mixed numbers is just a matter of first changing both of them to improper fractions and then invert and multiply.

Example:

$$4\frac{1}{3} \div 2\frac{3}{4} = \frac{13}{3} \div \frac{11}{4} = \frac{13}{3} \times \frac{4}{11} = \frac{52}{33} = 1\frac{19}{33}$$

You do these:

1. $6\frac{2}{5} \div 1\frac{1}{2}$

2. $5\frac{2}{3} \div 2\frac{1}{8}$

Answers

1. $6\frac{2}{5} \div 1\frac{1}{2} = \frac{32}{5} \div \frac{3}{2} = \frac{32}{5} \times \frac{2}{3} = \frac{64}{15} = 4\frac{4}{15}$

2. $5\frac{2}{3} \div 2\frac{1}{8} = \frac{17}{3} \div \frac{17}{8} = \frac{17}{3} \times \frac{8}{17} = \frac{136}{51} = \frac{8}{3} = 2\frac{2}{3}$

Steps for dividing fractions:

- Did you **change both mixed numbers** to improper fractions first?
- Did you **invert** the number being divided **by** and change to muliplication?
- Did you **multiply numerators** to get a new numerator?
- Did you **multiply denominators** to get a new denominator?
- Did you **reduce** your answer?
- Did you **change the improper fraction** in the answer to a mixed number?

IF YOU CAN REMEMBER TO DO THESE STEPS, YOU CAN DIVIDE FRACTIONS!

Practice Problems

1. $3\frac{1}{4} \div 2\frac{3}{5}$ 2. $6\frac{1}{2} \div 2\frac{1}{3}$ 3. $5\frac{2}{7} \div 2\frac{2}{3}$ 4. $6\frac{1}{2} \div 3\frac{7}{8}$

5. $9\frac{2}{3} \div 5\frac{1}{2}$ 6. $6\frac{3}{4} \div 1\frac{1}{8}$ 7. $8\frac{1}{4} \div 2\frac{1}{8}$ 8. $6\frac{7}{8} \div 2\frac{1}{4}$

9. $10\frac{1}{9} \div 1\frac{2}{3}$ 10. $8\frac{1}{5} \div 6\frac{1}{4}$ 11. $3\frac{1}{9} \div 3\frac{1}{3}$ 12. $8\frac{1}{8} \div 4\frac{1}{16}$

13. $6\frac{3}{5} \div 1\frac{3}{4}$ 14. $11\frac{1}{3} \div 4\frac{2}{5}$ 15. $6\frac{1}{9} \div 1\frac{7}{8}$ 16. $2\frac{1}{3} \div 5\frac{5}{8}$

17. $3\frac{1}{9} \div 6\frac{3}{5}$ 18. $4\frac{31}{49} \div 1\frac{1}{7}$ 19. $5\frac{2}{7} \div 2\frac{1}{4}$ 20. $6\frac{6}{7} \div 1\frac{1}{3}$

21. $5\frac{2}{3} \div 2\frac{3}{4}$ 22. $7\frac{1}{4} \div 3\frac{1}{2}$ 23. $3\frac{1}{8} \div 2\frac{1}{4}$ 24. $8\frac{2}{3} \div 6\frac{1}{2}$

25. $4\frac{3}{4} \div 8\frac{1}{3}$ 26. $15\frac{1}{2} \div 2\frac{1}{4}$ 27. $8\frac{3}{8} \div 3\frac{1}{5}$ 28. $6\frac{6}{7} \div 1\frac{3}{14}$

29. $1\frac{1}{5} \div 2\frac{2}{5}$ 30. $11\frac{1}{10} \div 2\frac{1}{5}$ 31. $4\frac{5}{6} \div 1\frac{1}{2}$ 32. $3\frac{1}{4} \div 1\frac{3}{8}$

33. $5\frac{2}{7} \div 2\frac{2}{3}$ 34. $6\frac{1}{2} \div 3\frac{7}{8}$ 35. $9\frac{2}{3} \div 5\frac{1}{2}$ 36. $6\frac{3}{4} \div 1\frac{1}{8}$

37. $8\frac{1}{4} \div 2\frac{1}{8}$ 38. $6\frac{7}{8} \div 2\frac{1}{4}$ 39. $10\frac{1}{9} \div 1\frac{2}{3}$ 40. $8\frac{1}{5} \div 6\frac{1}{4}$

41. $3\frac{1}{9} \div 3\frac{1}{3}$ 42. $8\frac{3}{4} \div 2\frac{7}{8}$ 43. $6\frac{3}{4} \div 1\frac{1}{2}$ 44. $3\frac{3}{5} \div 1\frac{1}{4}$

45. $9\frac{1}{10} \div 3\frac{3}{5}$ 46. $4\frac{1}{3} \div 6\frac{2}{5}$ 47. $4\frac{3}{7} \div 1\frac{5}{6}$ 48. $5\frac{1}{2} \div 3\frac{3}{4}$

49. $7\frac{7}{10} \div 2\frac{1}{8}$ 50. $1\frac{1}{4} \div 2\frac{2}{3}$ 51. $7\frac{1}{3} \div 4\frac{1}{2}$ 52. $5\frac{3}{4} \div 8\frac{1}{2}$

53. $12\frac{1}{3} \div 2\frac{2}{9}$ 54. $6\frac{6}{7} \div 1\frac{9}{14}$ 55. $3\frac{3}{7} \div 1\frac{1}{6}$ 56. $3\frac{1}{2} \div 2\frac{1}{2}$

Practice Problems

Reduce answers:

1. $\frac{1}{2} \div \frac{3}{4}$ 2. $\frac{4}{7} \div \frac{1}{3}$ 3. $\frac{7}{8} \div \frac{2}{5}$ 4. $\frac{6}{11} \div \frac{1}{4}$

5. $\frac{9}{10} \div \frac{1}{5}$ 6. $\frac{3}{4} \div \frac{1}{4}$ 7. $\frac{7}{9} \div \frac{2}{3}$ 8. $\frac{6}{7} \div \frac{3}{4}$

9. $\frac{1}{15} \div \frac{1}{5}$ 10. $\frac{3}{4} \div \frac{1}{8}$ 11. $\frac{5}{6} \div \frac{1}{2}$ 12. $\frac{5}{12} \div \frac{2}{3}$

13. $6 \div \frac{3}{7}$ 14. $4 \div \frac{3}{4}$ 15. $12 \div \frac{5}{6}$ 16. $9 \div \frac{7}{8}$

17. $15 \div \frac{1}{3}$ 18. $26 \div \frac{1}{9}$ 19. $16 \div \frac{4}{9}$ 20. $18 \div \frac{3}{7}$

21. $6\frac{1}{2} \div \frac{1}{3}$ 22. $4\frac{1}{3} \div \frac{3}{4}$ 23. $7\frac{1}{8} \div \frac{5}{6}$ 24. $3\frac{1}{3} \div \frac{1}{6}$

25. $24\frac{1}{4} \div \frac{3}{4}$ 26. $9\frac{1}{7} \div \frac{7}{9}$ 27. $6\frac{3}{4} \div \frac{4}{9}$ 28. $5\frac{5}{8} \div \frac{2}{5}$

29. $4\frac{1}{4} \div 3$ 30. $6\frac{2}{3} \div 2$ 31. $8\frac{4}{5} \div 3$ 32. $12\frac{3}{8} \div 4$

33. $3\frac{1}{3} \div 1\frac{1}{4}$ 34. $8\frac{1}{4} \div 2\frac{2}{3}$ 35. $7\frac{1}{2} \div 2\frac{1}{4}$ 36. $5\frac{1}{9} \div 1\frac{3}{8}$

37. $6\frac{1}{4} \div 3\frac{1}{8}$ 38. $9\frac{9}{10} \div 3\frac{1}{5}$ 39. $4\frac{1}{9} \div 6\frac{2}{3}$ 40. $3\frac{3}{5} \div 6\frac{1}{2}$

Practice Problems

Reduce answers:

1. $\frac{4}{5} \div \frac{1}{4}$ 2. $\frac{5}{6} \div \frac{2}{3}$ 3. $\frac{5}{8} \div \frac{1}{2}$ 4. $\frac{7}{9} \div \frac{3}{4}$

5. $\frac{6}{7} \div \frac{1}{7}$ 6. $\frac{13}{15} \div \frac{1}{3}$ 7. $\frac{20}{21} \div \frac{4}{5}$ 8. $\frac{9}{10} \div \frac{5}{8}$

9. $\frac{3}{5} \div \frac{1}{6}$ 10. $\frac{3}{10} \div \frac{3}{7}$ 11. $\frac{7}{9} \div \frac{1}{3}$ 12. $\frac{7}{12} \div \frac{5}{6}$

13. $4 \div \frac{1}{4}$ 14. $9 \div \frac{2}{3}$ 15. $7 \div \frac{7}{8}$ 16. $12 \div \frac{1}{3}$

17. $28 \div \frac{2}{3}$ 18. $63 \div \frac{9}{10}$ 19. $13 \div \frac{4}{7}$ 20. $21 \div \frac{5}{6}$

21. $4\frac{2}{5} \div \frac{1}{4}$ 22. $5\frac{2}{9} \div \frac{2}{3}$ 23. $7\frac{1}{2} \div \frac{1}{6}$ 24. $9\frac{1}{5} \div \frac{3}{8}$

25. $17\frac{1}{5} \div \frac{2}{5}$ 26. $6\frac{1}{4} \div \frac{1}{6}$ 27. $21\frac{1}{2} \div \frac{1}{4}$ 28. $48\frac{1}{8} \div \frac{3}{4}$

29. $6\frac{1}{3} \div 2$ 30. $8\frac{1}{5} \div 12$ 31. $14\frac{4}{5} \div 3$ 32. $26\frac{1}{4} \div 6$

33. $7\frac{1}{4} \div 2\frac{1}{3}$ 34. $9\frac{2}{3} \div 2\frac{1}{2}$ 35. $16\frac{1}{5} \div 3\frac{4}{5}$ 36. $24\frac{1}{3} \div 1\frac{1}{3}$

37. $66\frac{1}{5} \div 8\frac{1}{2}$ 38. $14\frac{2}{5} \div 6\frac{2}{5}$ 39. $70\frac{7}{8} \div 6\frac{1}{4}$ 40. $46\frac{2}{5} \div 5\frac{1}{10}$

Using Division of Fractions

1. Three farmers split $2\frac{1}{2}$ tons of hay evenly among themselves. How many tons of hay does each farmer get?

2. The $7\frac{1}{2}$ foot board was cut into four equal pieces. How long was each piece cut?

3. An angle is $31\frac{1}{2}$ degrees and is divided into two angles. How many degrees would be in each of the angles?

4. Sandy buys $4\frac{3}{4}$ yards of material. If she cuts the material into two equal pieces, how long is each piece of material?

5. A restaurant's inventory shows that they use $520\frac{1}{2}$ pounds of hamburger in one week. On the average, how many pounds of hamburger did the restaurant use per day? [Use a seven day work week.]

Using Division of Fractions (cont.)

6. Belinda cut a piece of ribbon into four equal pieces. If the original ribbon was $62\frac{1}{2}$ inches long, how long was one of the cut pieces?

7. A $6\frac{3}{4}$ inch strip of paper is cut into $\frac{3}{4}$ inch strips. How many $\frac{3}{4}$ inch strips can you get from the larger strip?

8. An $8\frac{1}{4}$ gallon barrel of water must be divided into small jars of $\frac{3}{8}$ gallon of water. How many small jars of water can you get from the barrel?

9. A cookie company has 300 pounds of butter in stock. One batch of cookies requires $\frac{3}{4}$ of a pound of butter. How many batches of cookies can the company make with the butter in stock?

10. Randall has a $10\frac{1}{2}$ inch strip of leather. How many $2\frac{5}{8}$ inch strips can he get from this larger strip?

Congratulations!

You have mastered fractions!!!

Check your Understanding

Use the fraction $\frac{3}{4}$ to answer problems 1 and 2.

1. Is "3" the numerator or denominator?

2. Is "4" the numerator or denominator?

In problems 3, 4, and 5, write down if they are a proper fraction, improper fraction, or a mixed number.

3. $\frac{8}{3}$ 4. $4\frac{1}{5}$ 5. $\frac{7}{8}$

6. What is $\frac{32}{32}$ equal to?

Reduce the following:

7. $\frac{18}{48}$ 8. $\frac{15}{27}$

Change the following to mixed numbers:

9. $\frac{17}{3}$ 10. $\frac{29}{6}$

Change the following to improper fractions: 11. $7\frac{2}{5}$ 12. $4\frac{3}{8}$

13. $\frac{2}{5} + \frac{1}{5}$

14. What is a common denominator of these two fractions? $\frac{1}{8}$ and $\frac{5}{12}$

15. $\frac{3}{8} + \frac{2}{3}$ 16. $\frac{5}{9} + \frac{3}{4}$ 17. $8 + 6\frac{1}{3}$ 18. $7\frac{2}{9} + 1\frac{6}{7}$

19. $\dfrac{7}{15} - \dfrac{3}{15}$

20. $\dfrac{6}{7} - \dfrac{1}{4}$

21. $\dfrac{9}{14} - \dfrac{3}{7}$

22. $6\frac{5}{9} - 5\frac{1}{3}$

23. $9\frac{1}{4} - 2\frac{3}{4}$

24. $7\frac{2}{5} - 3\frac{3}{4}$

25. $5 - 2\frac{2}{7}$

26. $\dfrac{2}{3} \times \dfrac{1}{4}$

27. $\dfrac{4}{5} \times \dfrac{3}{5}$

28. $6 \times \dfrac{2}{7}$

29. $\dfrac{3}{8} \times 4$

30. $\dfrac{3}{4} \times 2\frac{1}{2}$

31. $7\frac{1}{3} \times \dfrac{2}{9}$

32. $5\frac{1}{2} \times 2\frac{3}{4}$

33. $4\frac{2}{5} \times 3\frac{1}{4}$

34. $\frac{3}{8} \div \frac{2}{3}$

35. $\frac{1}{4} \div \frac{3}{5}$

36. $8 \div \frac{2}{7}$

37. $1\frac{1}{2} \div \frac{1}{4}$

38. $6\frac{2}{5} \div 2$

39. $8\frac{1}{4} \div 3\frac{3}{8}$

40. $5\frac{1}{9} \div 2\frac{3}{4}$

Check Your Understanding (cont.)

41. Home mortgage interest rates went down from $9\frac{1}{4}$ percent to $7\frac{1}{2}$ percent. How much did the interest rate decrease?

42. How many yards of cloth are in 8 rolls of cloth with $14\frac{1}{2}$ yards per roll?

43. If a $35\frac{3}{8}$ inch board is cut in half, how long are the two cut pieces?

44. If Bill jogs $2\frac{1}{2}$ miles in the morning and $3\frac{3}{4}$ miles in the afternoon, how many miles has he jogged?

45. If a computer stock was selling at $15\frac{7}{8}$ and the stock went up $\frac{3}{4}$, how much would the stock now be worth?

Check Your Understanding (cont.)

46. How many $2\frac{1}{2}$ inch strips of ribbon can you get from a 40 inch roll of ribbon?

47. A recipe calls for $\frac{3}{4}$ cups of butter. How many cups of butter is needed if you triple the recipe?

48. A 42 inch piece of molding has $18\frac{1}{4}$ inches cut off. How long is the piece of molding that is left?

49. If the interest on a credit card is $9\frac{1}{2}$ percent plus prime, how much is the interest rate for the credit card if the prime rate is $7\frac{1}{4}$ percent?

50. If Marilyn walks $2\frac{1}{2}$ miles each day, seven days a week, how many miles does she walk in a week?

Answers To Problems

Page 5

1. $\frac{4}{9}$ 2. $\frac{3}{4}$ 3. $\frac{1}{3}$ 4. $\frac{3}{11}$ 5. $\frac{1}{2}$ 6. $\frac{2}{3}$ 7. $\frac{2}{3}$ 8. $\frac{2}{5}$ 9. $\frac{7}{9}$

10. $\frac{17}{28}$ 11. $\frac{5}{9}$ 12. $\frac{1}{8}$ 13. $\frac{5}{9}$ 14. $\frac{7}{33}$ 15. $\frac{31}{41}$ 16. $\frac{3}{5}$ 17. $\frac{3}{7}$

18. $\frac{3}{7}$ 19. $\frac{3}{4}$ 20. $\frac{6}{13}$ 21. $\frac{8}{11}$ 22. $\frac{1}{2}$ 23. $\frac{3}{5}$ 24. $\frac{1}{3}$ 25. $\frac{37}{56}$

26. $\frac{5}{7}$ 27. $\frac{2}{3}$ 28. $\frac{3}{5}$ 29. $\frac{1}{3}$ 30. $\frac{3}{5}$ 31. $\frac{36}{49}$ 32. $\frac{21}{31}$

Page 8

1. $1\frac{1}{6}$ 2. $3\frac{2}{3}$ 3. $4\frac{1}{5}$ 4. 7 5. $3\frac{6}{7}$ 6. $3\frac{1}{3}$ 7. $3\frac{11}{18}$ 8. $3\frac{4}{9}$

9. $3\frac{1}{4}$ 10. $3\frac{3}{5}$ 11. $3\frac{3}{7}$ 12. $4\frac{7}{13}$ 13. $6\frac{2}{5}$ 14. $4\frac{1}{9}$ 15. $4\frac{3}{4}$

16. $2\frac{7}{15}$ 17. $2\frac{9}{20}$ 18. $1\frac{5}{8}$ 19. $2\frac{31}{40}$ 20. $4\frac{2}{5}$ 21. $3\frac{1}{2}$ 22. $1\frac{3}{5}$

23. $11\frac{1}{3}$ 24. $4\frac{1}{13}$ 25. $12\frac{11}{17}$ 26. $1\frac{29}{30}$ 27. $3\frac{2}{5}$ 28. $4\frac{67}{82}$ 29. $8\frac{1}{9}$

30. $11\frac{1}{2}$ 31. $1\frac{2}{3}$ 32. $2\frac{2}{9}$ 33. $6\frac{7}{9}$ 34. $7\frac{1}{3}$ 35. $17\frac{1}{4}$ 36. $4\frac{39}{50}$

37. $4\frac{39}{61}$ 38. 6 39. $18\frac{1}{4}$ 40. $6\frac{2}{7}$ 41. $15\frac{1}{3}$ 42. $5\frac{1}{5}$ 43. $4\frac{7}{18}$

44. $3\frac{1}{20}$ 45. $6\frac{21}{23}$ 46. $3\frac{2}{15}$ 47. $2\frac{1}{2}$ 48. $2\frac{1}{4}$ 49. $32\frac{4}{31}$ 50. $16\frac{31}{49}$

Page 11

1. $\frac{11}{8}$ 2. $\frac{21}{5}$ 3. $\frac{17}{3}$ 4. $\frac{29}{4}$ 5. $\frac{24}{7}$ 6. $\frac{17}{2}$ 7. $\frac{23}{4}$ 8. $\frac{61}{5}$

9. $\frac{19}{10}$ 10. $\frac{45}{7}$ 11. $\frac{113}{8}$ 12. $\frac{93}{10}$ 13. $\frac{35}{12}$ 14. $\frac{92}{3}$ 15. $\frac{113}{13}$

16. $\frac{44}{9}$ 17. $\frac{27}{5}$ 18. $\frac{212}{5}$ 19. $\frac{147}{20}$ 20. $\frac{31}{9}$ 21. $\frac{157}{3}$ 22. $\frac{155}{8}$

23. $\frac{37}{4}$ 24. $\frac{323}{12}$ 25. $\frac{83}{8}$ 26. $\frac{19}{4}$ 27. $\frac{92}{7}$ 28. $\frac{19}{2}$ 29. $\frac{44}{3}$

30. $\frac{93}{14}$ 31. $\frac{87}{10}$ 32. $\frac{23}{7}$ 33. $\frac{243}{5}$ 34. $\frac{103}{10}$ 35. $\frac{166}{29}$ 36. $\frac{121}{8}$

37. $\frac{14}{11}$ 38. $\frac{42}{17}$ 39. $\frac{71}{8}$ 40. $\frac{23}{5}$ 41. $\frac{145}{16}$ 42. $\frac{460}{9}$ 43. $\frac{137}{9}$

44. $\frac{81}{13}$ 45. $\frac{39}{10}$ 46. $\frac{31}{15}$ 47. $\frac{169}{21}$ 48. $\frac{58}{17}$ 49. $\frac{515}{22}$ 50. $\frac{579}{8}$

Page 12

1. Proper 2. Mixed 3. Proper 4. Improper 5. Mixed
6. Improper 7. Proper 8. Improper 9. Mixed 10. Proper

11. $\frac{4}{5}$ 12. $\frac{2}{3}$ 13. $\frac{2}{9}$ 14. $\frac{1}{2}$ 15. $\frac{1}{15}$

16. $\frac{2}{3}$ 17. $\frac{8}{9}$ 18. $\frac{12}{17}$ 19. $\frac{1}{3}$ 20. $\frac{29}{30}$

Page 12 (cont.)

21. $3\frac{1}{4}$ 22. $5\frac{2}{5}$ 23. $3\frac{2}{13}$ 24. $7\frac{2}{9}$ 25. $2\frac{5}{7}$

26. $16\frac{1}{3}$ 27. $1\frac{2}{7}$ 28. $4\frac{1}{5}$ 29. $5\frac{11}{20}$ 30. $3\frac{1}{2}$

31. $\frac{47}{8}$ 32. $\frac{7}{2}$ 33. $\frac{48}{5}$ 34. $\frac{11}{3}$ 35. $\frac{97}{6}$

36. $\frac{13}{7}$ 37. $\frac{113}{9}$ 38. $\frac{19}{4}$ 39. $\frac{25}{3}$ 40. $\frac{21}{8}$

Page 13

1. Mixed 2. Improper 3. Improper 4. Proper 5. Proper
6. Improper 7. Mixed 8. Proper 9. Mixed 10. Improper

11. $\frac{6}{7}$ 12. $\frac{2}{3}$ 13. $\frac{1}{4}$ 14. $\frac{1}{3}$ 15. $\frac{4}{9}$

16. $\frac{12}{25}$ 17. $\frac{43}{64}$ 18. $\frac{1}{3}$ 19. $\frac{12}{13}$ 20. $\frac{73}{109}$

21. $1\frac{8}{9}$ 22. $12\frac{1}{5}$ 23. $7\frac{5}{6}$ 24. $4\frac{2}{3}$ 25. $36\frac{1}{2}$

26. $2\frac{2}{7}$ 27. $3\frac{1}{12}$ 28. $9\frac{1}{2}$ 29. $1\frac{23}{30}$ 30. $3\frac{13}{20}$

31. $\frac{29}{8}$ 32. $\frac{57}{8}$ 33. $\frac{19}{2}$ 34. $\frac{93}{5}$ 35. $\frac{16}{9}$

36. $\frac{137}{3}$ 37. $\frac{67}{5}$ 38. $\frac{41}{10}$ 39. $\frac{221}{100}$ 40. $\frac{181}{35}$

Page 17

1. $\frac{4}{5}$ 2. $\frac{6}{7}$ 3. $1\frac{1}{2}$ 4. $\frac{8}{11}$ 5. $\frac{23}{25}$ 6. $1\frac{2}{9}$ 7. $\frac{7}{9}$ 8. $1\frac{2}{7}$

9. $\frac{12}{19}$ 10. $\frac{22}{23}$ 11. $\frac{37}{41}$ 12. 1 13. $\frac{6}{7}$ 14. $\frac{23}{31}$ 15. $\frac{12}{25}$

16. $\frac{40}{47}$ 17. $\frac{5}{11}$ 18. $1\frac{1}{3}$ 19. $\frac{46}{53}$ 20. $\frac{9}{11}$ 21. $\frac{114}{115}$ 22. $1\frac{2}{37}$

23. $\frac{25}{47}$ 24. $\frac{16}{19}$

Page 23

1. $\frac{13}{15}$ 2. $\frac{15}{28}$ 3. $1\frac{1}{9}$ 4. $1\frac{1}{8}$ 5. $1\frac{5}{14}$ 6. $\frac{7}{12}$ 7. $\frac{19}{45}$ 8. $1\frac{5}{18}$

9. $1\frac{1}{40}$ 10. $1\frac{1}{4}$ 11. $1\frac{5}{24}$ 12. $1\frac{1}{3}$ 13. $1\frac{7}{15}$ 14. $\frac{15}{56}$ 15. $\frac{7}{18}$

16. $\frac{13}{21}$ 17. $\frac{29}{30}$ 18. $\frac{7}{10}$ 19. $\frac{35}{39}$ 20. $\frac{25}{72}$ 21. $\frac{2}{3}$ 22. $\frac{65}{66}$

23. $\frac{13}{30}$ 24. $\frac{5}{6}$ 25. $\frac{49}{60}$ 26. $\frac{90}{217}$ 27. $\frac{69}{82}$ 28. $\frac{23}{36}$ 29. $1\frac{9}{170}$

30. $\frac{5}{12}$

Page 24

1. $\frac{22}{35}$ 2. $\frac{17}{24}$ 3. $\frac{31}{38}$ 4. $\frac{3}{4}$ 5. $1\frac{1}{24}$ 6. $\frac{46}{65}$ 7. $\frac{7}{9}$ 8. $1\frac{1}{12}$

9. $\frac{1}{3}$ 10. $\frac{3}{5}$ 11. $\frac{3}{5}$ 12. $\frac{17}{18}$ 13. $1\frac{7}{15}$ 14. $\frac{13}{24}$ 15. $\frac{31}{99}$

16. $1\frac{1}{10}$ 17. $\frac{31}{56}$ 18. $\frac{7}{15}$ 19. $\frac{41}{52}$ 20. $\frac{23}{30}$ 21. $\frac{79}{129}$ 22. $\frac{61}{63}$

23. $1\frac{7}{24}$ 24. $\frac{7}{16}$ 25. $\frac{41}{68}$ 26. $\frac{29}{45}$ 27. $\frac{35}{72}$ 28. $\frac{17}{30}$ 29. $1\frac{38}{77}$

30. $\frac{57}{100}$ 31. $1\frac{13}{24}$ 32. $1\frac{4}{15}$ 33. $1\frac{7}{18}$

Page 26

1. $3\frac{3}{4}$ 2. $8\frac{1}{3}$ 3. $3\frac{8}{15}$ 4. $4\frac{7}{18}$ 5. $11\frac{41}{44}$ 6. $6\frac{1}{12}$ 7. $5\frac{7}{10}$

8. $15\frac{23}{45}$ 9. $9\frac{11}{28}$ 10. $11\frac{3}{16}$ 11. $6\frac{7}{72}$ 12. $14\frac{11}{15}$

Page 27

1. $5\frac{12}{35}$ 2. $5\frac{19}{24}$ 3. $6\frac{25}{63}$ 4. $11\frac{21}{55}$ 5. $6\frac{7}{15}$ 6. $12\frac{5}{6}$ 7. $11\frac{2}{5}$ 8. $11\frac{1}{4}$

9. $8\frac{7}{9}$ 10. $12\frac{3}{10}$ 11. $23\frac{11}{15}$ 12. $6\frac{61}{76}$ 13. $13\frac{19}{20}$ 14. $16\frac{1}{10}$ 15. $17\frac{27}{70}$

16. $59\frac{7}{12}$ 17. $7\frac{23}{42}$ 18. $13\frac{5}{6}$ 19. $37\frac{7}{9}$ 20. $9\frac{11}{15}$ 21. $17\frac{17}{40}$ 22. $9\frac{23}{40}$

23. $9\frac{1}{18}$ 24. $7\frac{23}{44}$ 25. $25\frac{3}{4}$ 26. $46\frac{11}{12}$ 27. $37\frac{2}{3}$ 28. $21\frac{1}{6}$ 29. $10\frac{33}{70}$

30. $8\frac{61}{170}$ 31. $6\frac{7}{72}$ 32. $14\frac{11}{15}$ 33. $11\frac{1}{10}$ 34. $10\frac{19}{28}$ 35. $13\frac{4}{7}$

Page 28

1. $\frac{3}{4}$ 2. $1\frac{2}{5}$ 3. 1 4. $1\frac{1}{4}$ 5. $\frac{8}{11}$

6. $1\frac{4}{15}$ 7. $1\frac{17}{28}$ 8. $\frac{5}{8}$ 9. $1\frac{13}{24}$ 10. $1\frac{5}{26}$

11. $\frac{23}{60}$ 12. $\frac{29}{33}$ 13. $\frac{7}{32}$ 14. $\frac{9}{20}$ 15. $\frac{67}{100}$

16. $6\frac{1}{2}$ 17. $14\frac{2}{3}$ 18. $13\frac{7}{8}$ 19. $20\frac{5}{8}$ 20. $30\frac{1}{9}$

21. $5\frac{9}{20}$ 22. $7\frac{1}{6}$ 23. $8\frac{29}{40}$ 24. $14\frac{1}{18}$ 25. $5\frac{7}{30}$

26. $24\frac{5}{8}$ 27. $6\frac{9}{14}$ 28. $9\frac{1}{9}$ 29. $26\frac{5}{12}$ 30. $31\frac{1}{24}$

Page 29

1. 1 2. $\frac{7}{10}$ 3. $\frac{4}{5}$ 4. $1\frac{2}{7}$ 5. $\frac{12}{17}$

6. $\frac{41}{44}$ 7. $\frac{39}{56}$ 8. $\frac{40}{51}$ 9. $\frac{89}{90}$ 10. $\frac{33}{35}$

11. $\frac{71}{99}$ 12. $1\frac{1}{15}$ 13. $1\frac{7}{20}$ 14. $1\frac{7}{50}$ 15. $\frac{83}{90}$

16. $17\frac{7}{9}$ 17. $21\frac{2}{3}$ 18. $17\frac{1}{2}$ 19. $75\frac{11}{12}$ 20. $60\frac{11}{25}$

21. $7\frac{13}{14}$ 22. $8\frac{29}{45}$ 23. $13\frac{1}{8}$ 24. $38\frac{5}{9}$ 25. $24\frac{8}{15}$

26. $16\frac{11}{20}$ 27. $52\frac{15}{22}$ 28. $29\frac{9}{10}$ 29. $20\frac{55}{56}$ 30. $75\frac{11}{14}$

Pages 30-31

1. $3\frac{1}{6}$ miles 2. $2\frac{5}{8}$ carats 3. $1\frac{5}{6}$ cups 4. $107\frac{1}{8}$ inches 5. $14\frac{1}{4}$ hours

6. $48\frac{5}{8}$ inches 7. $17\frac{1}{2}$ pounds 8. $3\frac{1}{12}$ inches 9. $25\frac{11}{12}$ feet 10. $24\frac{1}{6}$ pages

Page 35

1. $\frac{1}{2}$ 2. $\frac{1}{4}$ 3. $\frac{2}{3}$ 4. $\frac{3}{11}$ 5. $\frac{4}{7}$ 6. $\frac{4}{9}$ 7. $\frac{1}{3}$ 8. $\frac{8}{17}$

9. $\frac{2}{3}$ 10. $\frac{1}{5}$ 11. $\frac{2}{5}$ 12. $\frac{4}{13}$ 13. $\frac{8}{17}$ 14. $\frac{3}{25}$ 15. $\frac{8}{31}$

16. $\frac{1}{4}$ 17. $\frac{1}{11}$ 18. $\frac{6}{23}$ 19. $\frac{5}{7}$ 20. $\frac{22}{47}$

Page 36

1. $\frac{1}{8}$ 2. $\frac{3}{28}$ 3. $\frac{2}{9}$ 4. $\frac{13}{55}$ 5. $\frac{19}{36}$ 6. $\frac{19}{33}$ 7. $\frac{9}{16}$ 8. $\frac{7}{12}$

9. $\frac{3}{10}$ 10. $\frac{29}{117}$ 11. $\frac{11}{21}$ 12. $\frac{9}{50}$ 13. $\frac{8}{35}$ 14. $\frac{3}{26}$

15. $\frac{7}{18}$ 16. $\frac{3}{28}$ 17. $\frac{1}{6}$ 18. $\frac{7}{20}$ 19. $\frac{3}{8}$ 20. $\frac{3}{16}$ 21. $\frac{3}{8}$

22. $\frac{19}{40}$ 23. $\frac{8}{45}$ 24. $\frac{3}{20}$ 25. $\frac{31}{63}$ 26. $\frac{1}{3}$ 27. $\frac{5}{122}$ 28. $\frac{53}{85}$

29. $\frac{19}{46}$ 30. $\frac{20}{63}$ 31. $\frac{1}{2}$ 32. $\frac{5}{24}$ 33. $\frac{1}{8}$ 34. $\frac{3}{32}$ 35. $\frac{5}{8}$

36. $\frac{3}{10}$ 37. $\frac{300}{559}$ 38. $\frac{61}{90}$ 39. $\frac{1}{21}$ 40. $\frac{1}{12}$ 41. $\frac{17}{111}$ 42. $\frac{37}{92}$

Page 40

1. $2\frac{1}{4}$ 2. $3\frac{3}{4}$ 3. $3\frac{7}{24}$ 4. $6\frac{1}{12}$ 5. $3\frac{19}{72}$ 6. $3\frac{3}{4}$ 7. $1\frac{17}{45}$ 8. $5\frac{1}{10}$

9. $3\frac{1}{3}$ 10. $\frac{7}{8}$ 11. $1\frac{1}{6}$ 12. $5\frac{5}{12}$ 13. $5\frac{19}{40}$ 14. $5\frac{5}{8}$ 15. $4\frac{8}{45}$

Page 40 (cont.)

16. $3\frac{3}{4}$ 17. $6\frac{3}{8}$ 18. $9\frac{17}{24}$ 19. $5\frac{3}{10}$ 20. $24\frac{7}{8}$ 21. $4\frac{4}{9}$ 22. $4\frac{23}{40}$

23. $2\frac{7}{9}$ 24. $6\frac{5}{9}$ 25. $11\frac{5}{6}$ 26. $55\frac{19}{28}$ 27. $8\frac{1}{24}$ 28. $1\frac{14}{15}$ 29. $23\frac{3}{8}$

30. $9\frac{19}{45}$

Page 41

1. $5\frac{1}{30}$ 2. $4\frac{3}{20}$ 3. $2\frac{5}{8}$ 4. $3\frac{11}{40}$ 5. $2\frac{1}{2}$ 6. $3\frac{1}{9}$ 7. $\frac{1}{2}$ 8. $3\frac{13}{24}$

9. $4\frac{7}{12}$ 10. $14\frac{29}{40}$ 11. $4\frac{16}{35}$ 12. $2\frac{17}{63}$ 13. $5\frac{8}{9}$ 14. $3\frac{3}{8}$ 15. $6\frac{11}{56}$

16. $5\frac{4}{15}$ 17. $\frac{1}{3}$ 18. $3\frac{11}{15}$ 19. $3\frac{23}{30}$ 20. $4\frac{9}{44}$ 21. $4\frac{3}{4}$ 22. $2\frac{23}{24}$

23. $4\frac{11}{18}$ 24. $12\frac{7}{12}$ 25. $3\frac{7}{9}$ 26. $5\frac{1}{24}$ 27. $8\frac{5}{16}$ 28. $12\frac{13}{16}$ 29. $1\frac{5}{8}$

30. $15\frac{13}{24}$

Page 42

1. $\frac{7}{15}$ 2. $\frac{5}{11}$ 3. $\frac{1}{5}$ 4. $\frac{1}{17}$ 5. $\frac{9}{35}$

6. $\frac{3}{8}$ 7. $\frac{5}{12}$ 8. $\frac{33}{56}$ 9. $\frac{43}{99}$ 10. $\frac{48}{133}$

11. $\frac{1}{3}$ 12. $\frac{47}{72}$ 13. $\frac{41}{56}$ 14. $\frac{7}{36}$ 15. $\frac{7}{45}$

16. $4\frac{7}{12}$ 17. $6\frac{3}{10}$ 18. $3\frac{5}{26}$ 19. $3\frac{17}{24}$ 20. $17\frac{7}{45}$

21. $48\frac{17}{30}$ 22. $5\frac{13}{44}$ 23. $4\frac{13}{20}$ 24. $4\frac{7}{24}$ 25. $26\frac{1}{6}$

26. $\frac{7}{8}$ 27. $1\frac{29}{40}$ 28. $7\frac{27}{35}$ 29. $2\frac{5}{8}$ 30. $10\frac{3}{5}$

Page 43

1. $\frac{2}{13}$ 2. $\frac{11}{29}$ 3. $\frac{6}{35}$ 4. $\frac{18}{33}$ 5. $\frac{5}{21}$

6. $\frac{1}{22}$ 7. $\frac{34}{65}$ 8. $\frac{1}{21}$ 9. $\frac{13}{18}$ 10. $\frac{1}{15}$

11. $\frac{5}{34}$ 12. $\frac{39}{140}$ 13. $\frac{11}{40}$ 14. $\frac{65}{84}$ 15. $\frac{49}{64}$

16. $6\frac{1}{12}$ 17. $2\frac{11}{40}$ 18. $1\frac{19}{63}$ 19. $7\frac{11}{21}$ 20. $11\frac{11}{34}$

21. $4\frac{11}{12}$ 22. $1\frac{5}{8}$ 23. $5\frac{3}{5}$ 24. $7\frac{19}{24}$ 25. $3\frac{37}{63}$

26. $\frac{19}{36}$ 27. $18\frac{53}{55}$ 28. $13\frac{1}{3}$ 29. $46\frac{4}{7}$ 30. $26\frac{7}{12}$

Page 44

1. $123\frac{1}{2}$ pounds 2. $2\frac{3}{4}$ gallons 3. $49\frac{5}{8}$ inches 4. $2\frac{3}{4}$ gallons 5. $1\frac{3}{4}$ yards

6. $2\frac{1}{4}$ dozen 7. $5\frac{1}{2}$ miles 8. $88\frac{1}{2}$ inches 9. $\frac{5}{8}$ gallon 10. $1\frac{3}{4}$ pounds

Page 47

1. $\frac{2}{15}$ 2. $\frac{3}{20}$ 3. $\frac{7}{48}$ 4. $\frac{14}{27}$ 5. $\frac{4}{63}$ 6. $\frac{15}{44}$ 7. $\frac{4}{21}$ 8. $\frac{3}{14}$

9. $\frac{7}{55}$ 10. $\frac{3}{16}$ 11. 6 12. $\frac{1}{27}$ 13. $\frac{1}{14}$ 14. $\frac{11}{30}$ 15. $\frac{3}{10}$ 16. $\frac{7}{24}$

17. $\frac{1}{3}$ 18. $\frac{7}{18}$ 19. $\frac{2}{21}$ 20. $\frac{3}{7}$

Page 48

1. $\frac{3}{10}$ 2. $\frac{4}{7}$ 3. $\frac{3}{32}$ 4. $\frac{8}{21}$ 5. $\frac{3}{20}$ 6. $\frac{7}{15}$ 7. $1\frac{7}{11}$ 8. $\frac{1}{10}$

9. $\frac{3}{305}$ 10. $\frac{1}{18}$ 11. 6 12. $\frac{1}{27}$ 13. $\frac{2}{3}$ 14. $\frac{3}{5}$ 15. $\frac{3}{14}$ 16. $3\frac{3}{5}$

17. $\frac{112}{153}$ 18. $\frac{3}{20}$ 19. $\frac{4}{21}$ 20. $\frac{12}{55}$ 21. $\frac{2}{7}$ 22. $\frac{36}{65}$ 23. $2\frac{1}{4}$ 24. $6\frac{2}{3}$

Page 48 (cont.)

25. $\frac{2}{9}$ 26. $\frac{1}{3}$ 27. $\frac{8}{11}$ 28. $5\frac{1}{4}$ 29. $\frac{24}{85}$ 30. $\frac{1}{12}$ 31. $\frac{7}{20}$ 32. $\frac{3}{16}$

33. $\frac{14}{27}$ 34. $\frac{1}{8}$ 35. $\frac{2}{15}$ 36. $\frac{6}{11}$ 37. $\frac{1}{30}$ 38. $\frac{12}{17}$ 39. $\frac{5}{162}$ 40. $\frac{7}{27}$

41. $\frac{11}{18}$ 42. $\frac{3}{16}$ 43. $\frac{4}{15}$ 44. $\frac{9}{49}$ 45. $\frac{1}{2}$ 46. $\frac{24}{35}$ 47. $\frac{1}{24}$ 48. $\frac{12}{35}$

Page 50

1. $\frac{9}{10}$ 2. $1\frac{13}{32}$ 3. 4 4. $2\frac{1}{10}$ 5. $\frac{22}{35}$ 6. $1\frac{5}{18}$ 7. $\frac{5}{9}$ 8. $3\frac{5}{12}$

9. $1\frac{1}{32}$ 10. $1\frac{9}{10}$ 11. $6\frac{11}{18}$ 12. $1\frac{17}{18}$ 13. $\frac{7}{10}$ 14. $1\frac{6}{7}$ 15. $3\frac{19}{24}$ 16. $\frac{5}{8}$

17. $3\frac{5}{6}$ 18. $1\frac{11}{14}$ 19. $3\frac{4}{15}$ 20. $2\frac{3}{5}$ 21. $1\frac{19}{36}$ 22. $2\frac{8}{45}$ 23. $1\frac{2}{5}$ 24. $4\frac{1}{14}$

Page 51

1. $\frac{17}{20}$ 2. $1\frac{11}{24}$ 3. $3\frac{5}{7}$ 4. $4\frac{13}{18}$ 5. $1\frac{13}{32}$ 6. $3\frac{1}{4}$ 7. $\frac{7}{24}$ 8. $\frac{20}{21}$

9. $3\frac{1}{3}$ 10. $5\frac{1}{18}$ 11. $1\frac{7}{44}$ 12. $4\frac{13}{20}$ 13. $\frac{11}{15}$ 14. $\frac{4}{7}$ 15. $\frac{15}{32}$ 16. $\frac{25}{27}$

17. 3 18. $2\frac{2}{15}$ 19. $2\frac{19}{36}$ 20. $2\frac{3}{8}$ 21. $4\frac{1}{2}$ 22. $6\frac{3}{5}$ 23. $\frac{8}{15}$ 24. $\frac{29}{80}$

25. $\frac{11}{15}$ 26. $\frac{5}{6}$ 27. $3\frac{2}{7}$ 28. $4\frac{4}{35}$ 29. $3\frac{3}{20}$ 30. $1\frac{22}{35}$ 31. $4\frac{1}{40}$ 32. $\frac{19}{60}$

33. $1\frac{2}{7}$ 34. $6\frac{1}{12}$ 35. $1\frac{19}{30}$ 36. $4\frac{3}{10}$ 37. 3 38. $2\frac{2}{5}$ 39. $6\frac{1}{4}$ 40. $7\frac{1}{7}$

41. $\frac{13}{16}$ 42. $1\frac{13}{50}$ 43. $\frac{17}{96}$ 44. $1\frac{11}{24}$ 45. $5\frac{17}{32}$ 46. $2\frac{1}{8}$ 47. 3 48. $1\frac{67}{80}$

Page 53

1. $8\frac{5}{9}$ 2. $7\frac{7}{8}$ 3. $13\frac{9}{32}$ 4. $2\frac{2}{3}$ 5. $9\frac{9}{10}$ 6. 20 7. $23\frac{3}{8}$ 8. $17\frac{5}{12}$

9. $17\frac{11}{12}$ 10. $33\frac{3}{10}$ 11. $12\frac{3}{20}$ 12. $7\frac{19}{32}$ 13. $10\frac{1}{9}$ 14. $5\frac{11}{14}$ 15. $7\frac{13}{16}$ 16. 35

Page 54

1. 6 2. $14\frac{4}{9}$ 3. $6\frac{11}{12}$ 4. $7\frac{7}{10}$ 5. $19\frac{17}{24}$ 6. $27\frac{3}{10}$ 7. $9\frac{5}{8}$ 8. $3\frac{31}{35}$

9. $10\frac{26}{27}$ 10. $7\frac{19}{32}$ 11. $26\frac{5}{6}$ 12. $8\frac{5}{8}$ 13. $11\frac{43}{54}$ 14. $12\frac{1}{2}$ 15. $29\frac{11}{16}$ 16. $12\frac{1}{12}$

17. $9\frac{51}{64}$ 18. $19\frac{4}{5}$ 19. $14\frac{15}{28}$ 20. $9\frac{5}{6}$ 21. $14\frac{2}{9}$ 22. $26\frac{7}{12}$ 23. $22\frac{11}{15}$ 24. $10\frac{9}{16}$

25. $5\frac{4}{9}$ 26. $20\frac{2}{5}$ 27. $19\frac{19}{24}$ 28. $9\frac{11}{21}$ 29. $7\frac{5}{7}$ 30. $8\frac{29}{50}$ 31. $12\frac{3}{4}$ 32. $16\frac{15}{32}$

33. $34\frac{1}{8}$ 34. $8\frac{5}{8}$ 35. $3\frac{1}{21}$ 36. $35\frac{5}{8}$ 37. $28\frac{1}{20}$ 38. $8\frac{8}{35}$ 39. $6\frac{37}{48}$ 40. $22\frac{11}{27}$

41. $15\frac{5}{48}$ 42. $11\frac{7}{22}$ 43. $21\frac{21}{25}$ 44. $14\frac{2}{27}$ 45. $19\frac{1}{2}$ 46. 20 47. $5\frac{11}{32}$ 48. $8\frac{7}{16}$

49. $21\frac{1}{9}$ 50. $3\frac{3}{25}$ 51. $28\frac{1}{20}$ 52. $21\frac{17}{20}$ 53. $4\frac{7}{12}$ 54. $14\frac{7}{12}$ 55. $9\frac{31}{40}$ 56. $12\frac{6}{7}$

Page 55

1. $\frac{1}{3}$ 2. $\frac{8}{49}$ 3. $\frac{2}{25}$ 4. $\frac{2}{11}$ 5. $\frac{5}{8}$ 6. $\frac{9}{28}$ 7. $\frac{1}{18}$ 8. $\frac{4}{7}$ 9. $\frac{5}{32}$ 10. $\frac{20}{51}$

11. $\frac{1}{26}$ 12. $\frac{1}{6}$ 13. 4 14. 8 15. $\frac{4}{5}$ 16. 1 17. 16 18. $2\frac{5}{8}$ 19. $5\frac{1}{7}$

20. $8\frac{8}{9}$ 21. 1 22. 2 23. $\frac{25}{72}$ 24. $10\frac{5}{64}$ 25. $1\frac{4}{5}$ 26. $\frac{11}{15}$ 27. $1\frac{23}{27}$

Page 55 (cont.)

28. $8\frac{1}{3}$ 29. $4\frac{1}{2}$ 30. $14\frac{14}{15}$ 31. $16\frac{2}{5}$ 32. $18\frac{4}{7}$ 33. $8\frac{9}{20}$ 34. $14\frac{11}{32}$

35. $13\frac{25}{48}$ 36. $19\frac{19}{72}$ 37. $5\frac{16}{25}$ 38. $21\frac{53}{99}$ 39. $102\frac{19}{32}$ 40. $120\frac{5}{6}$

Page 56

1. $\frac{1}{18}$ 2. $\frac{3}{11}$ 3. $\frac{3}{16}$ 4. $\frac{7}{45}$ 5. $\frac{7}{16}$

6. $\frac{21}{65}$ 7. $\frac{1}{5}$ 8. $\frac{2}{3}$ 9. $\frac{2}{3}$ 10. $\frac{8}{17}$

11. $\frac{28}{45}$ 12. $\frac{27}{44}$ 13. $2\frac{2}{3}$ 14. $1\frac{1}{2}$ 15. $4\frac{4}{7}$

16. 8 17. $1\frac{1}{11}$ 18. $10\frac{1}{2}$ 19. $2\frac{1}{2}$ 20. $13\frac{1}{3}$

21. $\frac{9}{10}$ 22. $2\frac{1}{4}$ 23. $2\frac{6}{7}$ 24. $3\frac{29}{32}$ 25. $2\frac{5}{6}$

26. $\frac{23}{36}$ 27. 1 28. $4\frac{1}{2}$ 29. 7 30. $14\frac{1}{6}$

31. 14 32. $7\frac{13}{21}$ 33. $57\frac{3}{8}$ 34. 10 35. $24\frac{11}{32}$

36. $7\frac{11}{35}$ 37. $8\frac{17}{18}$ 38. 10 39. $19\frac{3}{4}$ 40. $7\frac{5}{7}$

Pages 57 - 58

1. $47\frac{1}{2}$ hours 2. $96\frac{1}{4}$ inches 3. $38\frac{1}{2}$ yards 4. $24\frac{1}{2}$ miles 5. $10\frac{1}{2}$ miles

6. $\frac{1}{6}$ of the whole cake 7. $1\frac{3}{10}$ pounds 8. $16\frac{7}{8}$ feet 9. $1\frac{1}{3}$ sticks 10. 360 employees

Page 61

1. $\frac{5}{6}$ 2. $1\frac{1}{2}$ 3. $2\frac{1}{3}$ 4. $\frac{8}{27}$ 5. 2 6. 3 7. $1\frac{5}{16}$ 8. $1\frac{1}{7}$ 9. $2\frac{2}{9}$

10. $\frac{14}{15}$ 11. $2\frac{8}{11}$ 12. $3\frac{3}{5}$ 13. $1\frac{7}{18}$ 14. $\frac{20}{21}$ 15. $1\frac{13}{32}$ 16. $1\frac{1}{12}$ 17. $\frac{7}{10}$

Page 61 (cont.)

18. $2\frac{1}{10}$ 19. $1\frac{1}{5}$ 20. $2\frac{2}{19}$ 21. $\frac{5}{26}$ 22. $2\frac{5}{8}$ 23. $\frac{35}{36}$ 24. $1\frac{1}{17}$ 25. 2

Page 62

1. $2\frac{1}{7}$ 2. $4\frac{1}{2}$ 3. $2\frac{1}{3}$ 4. $\frac{4}{7}$ 5. $1\frac{1}{11}$ 6. $13\frac{1}{3}$ 7. $\frac{8}{9}$ 8. $2\frac{4}{5}$ 9. $\frac{15}{28}$ 10. $1\frac{3}{5}$

11. $\frac{3}{22}$ 12. $\frac{9}{32}$ 13. $10\frac{5}{8}$ 14. $\frac{3}{70}$ 15. $6\frac{6}{7}$ 16. $3\frac{3}{8}$ 17. $\frac{3}{26}$ 18. $\frac{18}{25}$ 19. $2\frac{5}{6}$

20. $\frac{10}{21}$ 21. $3\frac{9}{13}$ 22. $1\frac{7}{8}$ 23. $\frac{35}{36}$ 24. $\frac{7}{11}$ 25. $6\frac{2}{3}$ 26. $3\frac{1}{2}$ 27. $\frac{10}{63}$

28. $4\frac{4}{5}$ 29. $\frac{25}{28}$ 30. $\frac{1}{14}$ 31. 15 32. $\frac{3}{4}$ 33. $\frac{1}{54}$ 34. $2\frac{4}{5}$ 35. $\frac{7}{20}$ 36. $\frac{20}{27}$

37. $\frac{16}{19}$ 38. $4\frac{1}{5}$ 39. $\frac{3}{26}$ 40. 4 41. $\frac{7}{32}$ 42. $\frac{18}{19}$ 43. 18 44. $\frac{3}{22}$ 45. $\frac{5}{8}$

46. $2\frac{2}{11}$ 47. $\frac{10}{23}$ 48. $4\frac{9}{10}$ 49. 48 50. $6\frac{2}{5}$

Page 65

1. $6\frac{3}{4}$ 2. 21 3. $\frac{2}{5}$ 4. $9\frac{7}{12}$ 5. $8\frac{2}{3}$ 6. $8\frac{4}{9}$ 7. $8\frac{1}{3}$ 8. 14 9. $26\frac{2}{3}$ 10. $\frac{7}{18}$

11. $\frac{9}{40}$ 12. $9\frac{3}{10}$ 13. $\frac{18}{25}$ 14. $\frac{42}{143}$ 15. $7\frac{7}{8}$ 16. $\frac{8}{45}$ 17. $6\frac{1}{3}$ 18. $11\frac{11}{36}$ 19. $8\frac{3}{4}$

20. $\frac{1}{3}$ 21. 7 22. $5\frac{53}{63}$ 23. $8\frac{2}{3}$ 24. $8\frac{20}{27}$ 25. $\frac{48}{65}$ 26. $\frac{9}{35}$ 27. $7\frac{7}{32}$ 28. $3\frac{1}{30}$

29. 15 30. $9\frac{7}{8}$

Page 66

1. $9\frac{1}{16}$ 2. $31\frac{1}{2}$ 3. $\frac{28}{81}$ 4. $\frac{9}{14}$ 5. $7\frac{7}{9}$ 6. $10\frac{4}{7}$ 7. $6\frac{1}{4}$ 8. $2\frac{4}{7}$ 9. $8\frac{1}{3}$ 10. $\frac{9}{32}$

11. $10\frac{5}{8}$ 12. $\frac{21}{88}$ 13. $\frac{11}{39}$ 14. $\frac{5}{7}$ 15. $5\frac{2}{7}$ 16. $\frac{35}{72}$ 17. $9\frac{5}{9}$ 18. $12\frac{3}{14}$ 19. $\frac{65}{112}$

20. $7\frac{31}{32}$ 21. $5\frac{1}{4}$ 22. $8\frac{17}{21}$ 23. $9\frac{3}{10}$ 24. $\frac{7}{18}$ 25. $13\frac{13}{24}$ 26. 10 27. $\frac{27}{91}$

28. $6\frac{1}{2}$ 29. $7\frac{7}{10}$ 30. $\frac{5}{12}$ 31. $6\frac{3}{8}$ 32. $9\frac{13}{27}$ 33. $7\frac{7}{30}$ 34. $\frac{7}{12}$ 35. $7\frac{3}{7}$ 36. $\frac{2}{15}$

37. $25\frac{1}{5}$ 38. $\frac{27}{119}$ 39. $\frac{7}{12}$ 40. $12\frac{1}{2}$ 41. $3\frac{1}{3}$ 42. 32 43. $8\frac{1}{3}$ 44. $13\frac{17}{27}$ 45. $24\frac{8}{9}$

Page 68

1. $1\frac{1}{4}$ 2. $2\frac{11}{14}$ 3. $1\frac{55}{56}$ 4. $1\frac{21}{31}$ 5. $1\frac{25}{33}$ 6. 6 7. $3\frac{15}{17}$ 8. $3\frac{1}{18}$ 9. $6\frac{1}{15}$

10. $1\frac{39}{125}$ 11. $\frac{14}{15}$ 12. 2 13. $3\frac{27}{35}$ 14. $2\frac{19}{33}$ 15. $3\frac{7}{27}$ 16. $\frac{56}{135}$ 17. $\frac{140}{297}$

18. $4\frac{3}{56}$ 19. $2\frac{22}{63}$ 20. $5\frac{1}{7}$ 21. $2\frac{2}{33}$ 22. $2\frac{1}{14}$ 23. $1\frac{7}{18}$ 24. $1\frac{1}{3}$ 25. $\frac{57}{100}$

26. $6\frac{8}{9}$ 27. $2\frac{79}{128}$ 28. $5\frac{11}{17}$ 29. $\frac{1}{2}$ 30. $5\frac{1}{22}$ 31. $3\frac{2}{9}$ 32. $2\frac{4}{11}$ 33. $1\frac{55}{56}$

34. $1\frac{21}{31}$ 35. $1\frac{25}{33}$ 36. 6 37. $3\frac{15}{17}$ 38. $3\frac{1}{18}$ 39. $6\frac{1}{15}$ 40. $1\frac{39}{125}$ 41. $\frac{14}{15}$

42. $3\frac{1}{23}$ 43. $4\frac{1}{2}$ 44. $2\frac{22}{25}$ 45. $2\frac{19}{36}$ 46. $\frac{65}{96}$ 47. $2\frac{32}{77}$ 48. $1\frac{7}{15}$ 49. $3\frac{53}{85}$

50. $\frac{15}{32}$ 51. $1\frac{17}{27}$ 52. $\frac{23}{34}$ 53. $5\frac{11}{20}$ 54. $4\frac{4}{23}$ 55. $2\frac{46}{49}$ 56. $1\frac{2}{5}$

Page 69

1. $\frac{2}{3}$ 2. $1\frac{5}{7}$ 3. $2\frac{3}{16}$ 4. $2\frac{2}{11}$ 5. $4\frac{1}{2}$

6. 3 7. $1\frac{1}{6}$ 8. $1\frac{1}{7}$ 9. $\frac{1}{3}$ 10. 6

11. $1\frac{2}{3}$ 12. $\frac{5}{8}$ 13. 14 14. $5\frac{1}{3}$ 15. $14\frac{2}{5}$

16. $10\frac{2}{7}$ 17. 45 18. 234 19. 36 20. 42

21. $19\frac{1}{2}$ 22. $5\frac{7}{9}$ 23. $8\frac{11}{20}$ 24. 20 25. $32\frac{1}{3}$

26. $11\frac{37}{49}$ 27. $15\frac{3}{16}$ 28. $14\frac{1}{16}$ 29. $1\frac{5}{12}$ 30. $3\frac{1}{3}$

31. $2\frac{14}{15}$ 32. $3\frac{3}{32}$ 33. $2\frac{2}{3}$ 34. $3\frac{3}{32}$ 35. $3\frac{1}{3}$

36. $3\frac{71}{99}$ 37. 2 38. $3\frac{3}{32}$ 39. $\frac{37}{60}$ 40. $\frac{36}{65}$

Page 70

1. $3\frac{1}{5}$ 2. $1\frac{1}{4}$ 3. $1\frac{1}{4}$ 4. $1\frac{1}{27}$ 5. 6

6. $2\frac{3}{5}$ 7. $1\frac{4}{21}$ 8. $1\frac{11}{25}$ 9. $3\frac{3}{5}$ 10. $\frac{7}{10}$

11. $2\frac{1}{3}$ 12. $\frac{7}{10}$ 13. 16 14. $13\frac{1}{2}$ 15. 8

16. 36 17. 42 18. 70 19. $22\frac{3}{4}$ 20. $25\frac{1}{5}$

21. $17\frac{3}{5}$ 22. $7\frac{5}{6}$ 23. 45 24. $24\frac{8}{15}$ 25. 43

26. $37\frac{1}{2}$ 27. 86 28. $64\frac{1}{6}$ 29. $3\frac{1}{6}$ 30. $\frac{41}{60}$

31. $4\frac{14}{15}$ 32. $4\frac{3}{8}$ 33. $3\frac{3}{28}$ 34. $3\frac{13}{15}$ 35. $4\frac{5}{19}$

36. $18\frac{1}{4}$ 37. $7\frac{67}{85}$ 38. $2\frac{1}{4}$ 39. $11\frac{17}{50}$ 40. $9\frac{5}{51}$

Pages 71 - 72

1. $\frac{5}{6}$ ton 2. $1\frac{7}{8}$ feet 3. $15\frac{3}{4}$ degrees 4. $2\frac{3}{8}$ yards 5. $74\frac{5}{14}$ pounds

6. $15\frac{5}{8}$ inches 7. 9 strips 8. 22 jars 9. 400 batches 10. 4 strips

Answers to Check Your Understanding

1. Numerator 2. Denominator 3. Improper fraction 4. Mixed number

5. Proper fraction 6. 1 7. $\frac{3}{8}$ 8. $\frac{5}{9}$

9. $5\frac{2}{3}$ 10. $4\frac{5}{6}$ 11. $\frac{37}{5}$ 12. $\frac{35}{8}$

13. $\frac{3}{5}$ 14. Common denominators of 8 and 12 are 24, 48, and 96. You could have picked either one of these and have been correct.

15. $1\frac{1}{24}$ 16. $1\frac{11}{36}$ 17. $14\frac{1}{3}$ 18. $9\frac{5}{63}$

19. $\frac{4}{15}$ 20. $\frac{17}{28}$ 21. $\frac{3}{14}$ 22. $1\frac{2}{9}$

23. $6\frac{1}{2}$ 24. $3\frac{13}{20}$ 25. $2\frac{5}{7}$ 26. $\frac{1}{6}$

27. $\frac{12}{25}$ 28. $1\frac{5}{7}$ 29. $1\frac{1}{2}$ 30. $1\frac{7}{8}$

31. $1\frac{17}{27}$ 32. $15\frac{1}{8}$ 33. $14\frac{3}{10}$ 34. $\frac{9}{16}$

35. $\frac{5}{12}$ 36. 28 37. 6 38. $3\frac{1}{5}$

39. $2\frac{4}{9}$ 40. $1\frac{85}{99}$

41. $1\frac{3}{4}$ percent 42. 116 yards 43. $17\frac{11}{16}$ inches 44. $6\frac{1}{4}$ miles 45. $16\frac{5}{8}$

46. 16 strips 47. $2\frac{1}{4}$ cups 48. $23\frac{3}{4}$ inches 49. $16\frac{3}{4}$ percent 50. $17\frac{1}{2}$ miles